Practical Finance for Property Investment

Practical Finance for Property Investment provides readers with an introduction to the most fundamental concepts, principles, analytical methods, and tools useful for making investing and financing decisions regarding income-producing property. The book begins by considering how to value income-producing property by forecasting a property's cash flows and estimating appropriate discount rates. It then discusses how both debt and private equity are used as methods to finance a property's acquisition. The book provides a thorough discussion of the taxation of property income as well as how investors can quantify the risks to investing in property. The book concludes with important considerations for investors when their investment thesis does not come to fruition.

Practical Finance for Property Investment offers a unique and novel pedagogy by pairing each book chapter with an in-depth real-world case study, which forces readers to confront the occasional tensions between finance theory and property investment practice. The book is designed for investors and students interested in learning what finance theory implies about property investment.

Readers and Instructors can access electronic resources, including the spreadsheets used in the textbook, at the book's website: www.routledge.com/9780367333041

Craig Furfine is a Clinical Professor of Finance at the Kellogg School of Management at Northwestern University where he teaches classes in corporate finance and real estate finance. Furfine has published scholarly research on interbank markets, banking, and real estate finance and 18 case studies covering a wide range of topics on the latter.

Practical Finance for Property Investment

Craig Furfine

LONDON AND NEW YORK

First published 2020
by Routledge
2 Park Square, Milton Park, Abingdon, Oxon OX14 4RN

and by Routledge
52 Vanderbilt Avenue, New York, NY 10017

Routledge is an imprint of the Taylor & Francis Group, an informa business

© 2020 Craig Furfine

The right of Craig Furfine to be identified as author of this work has been asserted by him in accordance with sections 77 and 78 of the Copyright, Designs and Patents Act 1988.

All rights reserved. No part of this book may be reprinted or reproduced or utilised in any form or by any electronic, mechanical, or other means, now known or hereafter invented, including photocopying and recording, or in any information storage or retrieval system, without permission in writing from the publishers.

Trademark notice: Product or corporate names may be trademarks or registered trademarks, and are used only for identification and explanation without intent to infringe.

British Library Cataloguing-in-Publication Data
A catalogue record for this book is available from the British Library

Library of Congress Cataloging-in-Publication Data
Names: Furfine, Craig, author.
Title: Practical finance for property investment / Craig Furfine.
Description: Abingdon, Oxon ; New York, NY : Routledge, 2020. | Includes bibliographical references. | Identifiers: LCCN 2019029025 (print) | LCCN 2019029026 (ebook) | ISBN 9780367333034 (hardback) | ISBN 9780367333041 (paperback) | ISBN 9780429319112 (ebook)
Subjects: LCSH: Real estate investment. | Real estate investment—Finance.
Classification: LCC HD1382.5 .F87 2020(print) | LCC HD1382.5 (ebook) | DDC 332.63/24—dc23
LC record available at https://lccn.loc.gov/2019029025
LC ebook record available at https://lccn.loc.gov/2019029026

ISBN: 978-0-367-33303-4 (hbk)
ISBN: 978-0-367-33304-1 (pbk)
ISBN: 978-0-429-31911-2 (ebk)

Typeset in Times New Roman
by Apex CoVantage, LLC

Visit the eResources: www.routledge.com/9780367333041

To my loving wife,
Natalie,
for whom I am eternally grateful.

Contents

	Acknowledgements	viii
1	Leasing	1
2	Property valuation	19
3	Debt financing	49
4	Equity partnerships	78
5	The taxation of property investment	100
6	The quantification of risk	121
7	When things go wrong	145
	Index	165

Acknowledgements

I would like to thank my colleague Ravi Jagannathan, who encouraged me to write a book on real estate finance. I also appreciate the tremendous help received from previous Kellogg MBA students. In particular, Jessica Zaski '10 (Wildcat), Daniel Kamerling '11 and Sara Lo '12 (Golden Opportunity), Benjamin Engleman '13 and Ricardo Ikeda '13 (Tulaberry), Sam Schey '13 (Workouts), and Jacob Shapiro '15 (Tale of Two Properties) provided excellent research assistance on the cases used in this book. Raul Tamez '18 (property valuation), Colin Sullivan '17 (equity partnerships), and Dan Tarpey '15 (quantification of risk) provided vital research support for particular chapters. Finally, Priya Shah '19 and Alyssa Boente '19 provided invaluable feedback on early drafts of the manuscript.

1 Leasing

Introduction

Our discussion of property investment begins with a description of the typical components of a lease. This is because as an investor in commercial property, you want to know what cash flow the property will be able to generate during the time that you own the building. The most significant source of cash flow is the rent paid by a property's tenants, and contracted rent payments are specified in each tenant's lease.

Upon closer examination, however, a lease specifies much more than the periodic rent owed by a particular tenant. More generally, a lease is a legal agreement between the owner of a set of property rights, the lessor or landlord, and the temporary user of those same rights, the lessee or tenant. From the investor's perspective, not only does the lease specify the rent to be paid, but it also typically contains additional features that may also affect the value of the property. This chapter begins with an overview of these features. This is then followed by a discussion of how property investors should compare leases, both with respect to their explicit value as well as their implicit benefits to any property.

Overview of lease features

Generally speaking, all leases will contain some common characteristics. These are:

1. The space
2. The term
3. The rent
4. Expense sharing
5. Concessions
6. Options

The space

The lease will specify the space being granted to the tenant for use over the course of the lease. In an apartment building, the space might be *Apartment 413*. In a single-tenant industrial space, it might be *4205 Industrial Park Way*. In an office building, the space might be the *7th floor*. If 4 tenants in the office building share the 7th floor, each tenant might have a lease specifying their right to occupy a particular 4500-square-foot area on the 7th floor. It is important to recognize that certain spaces have unique features that may impact how attractive the space is to potential tenants. The location of the space can be important – not only with respect to the building's location but also with respect to the location of distinct

2 Leasing

locations within a single building. Retail space closer to the street may be more valuable than similar space on the interior of the same building. Space on higher floors with better views may command higher rents than space on lower levels.

The term

The lease will also specify the term during which the tenant has use of the space. For example, an apartment lease might last for 12 months. The term of a typical lease varies by property type. Hotels (implicitly) have very short leases – overnight! Office tenants often sign leases for 5 or more years, whereas specialized industrial warehouses may lease for 10 or more years at a time.

The term of a lease will influence property cash flows because the expiration of a lease indicates the likelihood that a landlord needs to grant concessions (discussed further in a following section) to a new tenant. Thus, all else equal, a longer lease is beneficial for cash flow. Put another way, a landlord may be willing to accept lower rent in exchange for a longer term to avoid the costs associated with getting the existing tenant to sign a new lease or finding a new tenant. The risk of getting a tenant to sign a new lease is also likely to be systematic, in that the ease with which one can lease a property is related to the state of the economy. Therefore, cash flow from a property with higher vacancy (and thus a greater need for leasing) is riskier than cash flow from a similar property with less vacancy.

The rent

Quotation

Rent can loosely be defined as money paid by a tenant to the landlord in exchange for the use of the space specified in the lease during the lease's term. The way rent is described varies with property type. The rent for residential property is typically quoted in a dollar amount per month. For example, the tenant in Apartment 413 commits to paying $2500 per month over the term of the lease. Retail, office, and industrial property, however, is typically quoted in dollars per square foot per year.

> Example 1.1: Suppose an industrial tenant is obligated to pay $68,750 per month to lease 75,000 square feet. This rent would typically be quoted at $11 (per square foot per year). This can be calculated directly as $68,750 × 12 ÷ 75,000 = $11.
>
> Example 1.2: Similarly, a retail tenant's lease might be quoted at $30 (per square foot per year), which would imply a store owner leasing 5000 square feet would be obligated to pay $12,500 per month ($30 × 5000 ÷ 12 = $12,500).

Variation over time

Leases may specify **flat rent**, which means that the rental rate is constant over the duration of the lease. Alternatively, the lease may specify **graduated rent**. Graduated rent means that the rental rate is scheduled to change at particular times during the life of the lease and that the amount to which the rent is adjusted is known and specified in the lease.

> Example 1.3: A 7-year lease with graduated rent might specify that rent in Years 1 through 3 is $20, rent in Years 4 through 6 would increase to $21.50, and that rent during the final year of the lease would be $22.

Alternatively, a lease may specify **indexed rent**. Indexed rent adjusts at a pre-specified frequency but to an amount that is only determined after the fact based on a movement in some observable price level.

> Example 1.4: A 7-year lease might specify adjustments to the rent in Year 4 and Year 7, where the adjustment amount is indexed to be 75% of the change in the consumer price index (CPI) since the last adjustment. Were this form of rental adjustment specified in a 10-year lease beginning at $20 with adjustments in Years 4 and 7, rent would increase to $21.20 in Year 4 if the CPI increased by 8% during the first three years ($20 × [1 + (75% × 8%)]) = $21.20.

A final type of rent variation within a lease is **percentage rent**. This type of rent specifies a base rent and an additional component of rent that is related to some aspect of the financial performance of the tenant.

> Example 1.5: For instance, a retail tenant's 2000-square-foot lease might specify a base rent of $25 plus 5% of gross revenue above a predetermined threshold. If that excess revenue were $80,000, then the total rent payable under the lease would be $54,000 = ($25 × 2000) + (5% × $80,000) for the year, or an equivalent of $27 (i.e. $54,000 ÷ 2000) per square foot.

Percentage rents incentivize landlords to undertake leasing strategies that serve to improve the tenant's business outcomes. Likewise, percentage rents allow tenants to hedge their business risks by allowing them to pay lower rents when business is poor in exchange for higher rents when business is good.

Expense sharing

Landlords face a number of expenses related to the operation of an investment property. Leases specify whether some or all of certain expenses are to be paid by the tenant. In a **gross lease**, the landlord is responsible for paying all of the expenses. In a **net lease**, some of the operating expenses of the building are paid for by the tenant. In a **triple net lease (NNN)**, virtually all of the operating expenses of the building are paid by the tenant.

In a gross lease, operating expenses are paid by the landlord. A net lease will allocate certain expenses to the tenant. In situations where the tenant is responsible for certain operating expenses, those expenses might include cleaning, repairs, maintenance, landscaping, electricity, water and sewer, security, real estate taxes, and property insurance. Other expenses are typically not reimbursable. These include leasing commissions (LCs), administrative overhead, capital expenditures, and management fees. The expenses subject to reimbursement by the tenant will be specified in the tenant's lease.

As an investor, you may be thinking, "When I own this property, I'm going to insist that my tenants have NNN leases." This is because you believe it is better for you if your tenant pays for the majority of a building's operating expenses. Realize, however, that your tenants will understand that if they sign a NNN lease, they will have to pay operating expenses and will therefore expect a lower rent. For example, a tenant may consider a lease with gross rent of $35 to be equivalent to a lease with NNN rent of $22 if operating expenses are $13. In practice, landlords typically don't strategically choose between a net rather than a gross lease. Often, the presence of gross or net leases may be driven by property characteristics.

4 Leasing

For instance, an older property may not have its individual tenant spaces metered for electricity separately, which would make it difficult to offer a net lease.

In some instances, leases will specify a sharing of expenses between landlord and tenant that is somewhat between the extremes of either a gross or NNN lease. For instance, a lease might specify gross rent with an **expense stop**. A lease with an expense stop specifies that tenants will pay for **reimbursable expenses** that exceed a certain level, called the "stop." This type of expense sharing is designed to protect the landlord against inflation in operating expenses, while at the same time giving tenants the incentive to keep operating costs down.

> Example 1.6: Suppose the sole tenant of a 100,000-square-foot office building has an expense stop of $5. As long as reimbursable operating expenses remain below $500,000 in a given year, the landlord would pay all of the expenses. If reimbursable expenses were $600,000, then the landlord would pay $500,000 ($5 per square foot), whereas the tenant would be responsible for the additional $100,000.

In a multi-tenant property, it is likely that expense stops vary across leases. Thus, the responsibility to reimburse the landlord for expenses would vary across tenants.

> Example 1.7: For example, consider the list of tenants in Table 1.1, who occupy a given building that incurs annual reimbursable expenses of $324,000. This comes to $4.32 per square foot (per year), calculated as $324,000 ÷ 75,000 = $4.32. Reimbursements are calculated for each tenant based on the level of expenses above each tenant's stop.

Concessions

Leases will specify any **concessions** given to a tenant, which reflect money spent or foregone by the landlord in order to attract or maintain tenants. A **rent concession** is a temporary rent reduction given to tenants by the landlord. For example, a landlord may offer an office tenant 5 months of free rent with the signing of a 5-year lease. Other concessions are **tenant improvement** allowances, or TIs. TIs are capital expenditures incurred by the landlord to install or update office layouts and fixtures to suit the design needs of a tenant. For example, suppose a law firm is negotiating to lease space formerly rented by an internet startup company. The startup business operated with large open spaces, but the law firm needs a number of private conference rooms and corner offices for the partners. The landlord might offer the law firm a $50 (per square foot) TI upon signing a lease to be used towards remodeling the space to its own specifications.

Students sometimes wonder why a landlord grants an upfront concession rather than simply reducing the level of rent. Is there any reason to prefer concessions as opposed to rent reductions? Some investors believe that upfront concessions are particularly valuable to cash-strapped potential tenants. Thus, rent concessions and TIs serve to alleviate the credit constraints of the tenants.

Table 1.1 Calculating expense reimbursements

Tenant	Square Feet	Expense Stop	Reimbursements
Technology firm	50,000	$4.00	($4.32 − $4.00) × 50,000 = $16,000
Accounting firm	20,000	$4.25	($4.32 − $4.25) × 20,000 = $1400
Coffee shop	5,000	$5.00	$0

Other investors believe that offering large concessions early in a lease term allows them to "advertise" higher rental rates, which they believe will allow them to sell the property at higher prices. This relies on potential buyers being ignorant of the fact that current owners may have manipulated lease rates by offering concessions in the past.

Options

Leases often provide flexibility to both tenants and landlords through the provision of options. A **renewal or extension option** gives the tenant the option to renew the lease for an additional period of time. For example, a 10-year lease might include two 5-year renewal options, which gives the tenant confidence that it can secure its current space beyond the initial term of the lease. An **expansion option** gives the tenant the right to expand its occupancy into adjacent spaces that may become available during the time of the lease. This is particularly valuable to a company that anticipates a growth in its space needs over time but doesn't want to commit to leasing excess space in advance of its current need. Leases may contain **cancellation options**, where the landlord has the right to terminate the lease prior to its expiration. One common reason that landlords may negotiate a cancellation option is to preserve their right to redevelop the property at a future date. Tenants, too, may be given a cancellation option. For example, retail tenants may have the right to exit a lease if one of their co-tenants – perhaps the mall anchor – vacates the property.

Options are valuable, and therefore any option given to a tenant is value lost by the landlord. For example, suppose a landlord offers a tenant a 10-year lease at $25, with a 5-year extension option at $30. The tenant will only exercise the option if, at the end of 10 years, the market rent is greater than $30. Thus, the cost of offering the extension option to the tenant has something to do with the likelihood that rents will be higher than $30 in 10 years and, if so, how much higher will they be. This is a complicated option pricing problem that is beyond the scope of this book. Intuitively, however, it is useful to understand that, all else equal, otherwise identical extension options are more costly to the landlord in markets where rents are more volatile. Note that even if the extension option gives the tenant the right to renew "at market rent," it is still potentially costly to the landlord. This is because the landlord potentially gives up the right to replace that particular tenant with one with a better risk profile.

Balancing the cost of options granted are their potential benefits. For instance, an expansion option might be a lease feature that allows you to convince a fast-growing business to move into your property. Thus, it may be optimal to "invest" in granting costly options to your tenants if they attract higher quality tenants or reduce turnover costs and risks by keeping existing tenants in your property longer.

Comparing leases

As a property investor, you will be faced with negotiating leases with your tenants. This chapter has described the multitude of dimensions in which lease contracts can vary. How do you decide what is best? In this section, we review some popular approaches and discuss their limitations.

Lease present value

Through a series of examples, we will demonstrate the calculation of **lease present value**, which is simply the net present value of the cash flows associated with a given lease. These

6 Leasing

examples will illustrate why maximizing the present value of cash flows within a lease is not, by itself, the optimal way to choose among various leases.

> **Example 1.8**: Suppose you are looking to lease a 1000-square-foot retail space and a tenant offers to pay annual rent of $15 per square foot per year. The tenant also requires a $10 per square foot TI and you have to pay a leasing broker 5% of total lease revenue as a commission in exchange for bringing you this tenant. We will have much more to say about discount rates in a real estate context in the following chapter, but for now further suppose that the appropriate discount rate on these cash flows is 10% per year.

Graphically, these promised cash flows can be shown on a time line as given in Figure 1.1.

In the first year, you (the investor) have to pay for a $10,000 TI expense as well as a $3750 broker commission ($3750 = 5% × total lease revenue = 5% × 5 years × 1000 square feet × $15).[1] In exchange, you receive the first year's rent of $15,000 as well as an additional $15,000 in each of the subsequent 4 years. With a 10% discount rate, the present value of this lease's cash flows can be calculated to be $44,361.80.

> **Example 1.9**: You may have the intuition that the higher the NPV of a lease, the better it is for you, the investor. For example, if the tenant in the previous example were to agree to pay you $20 in rent, then holding all other terms of the lease constant the net present value of the lease would rise to $62,179.37. Clearly, all else equal, receiving higher rent would be better for the landlord/investor.
>
> **Example 1.10**: To begin to understand why net present value isn't the only consideration, imagine that the tenant instead was willing to sign a 50-year lease but only pay you $12 in rent per square foot per year. This lease would have an NPV of $82,614.14, which is noticeably higher than the present values of the other 2 leases. As the property owner, would this be better? Hopefully you can begin to see why NPV is not the sole determining factor of which lease to prefer. That is because the term of the lease (among other factors discussed) is also extremely important.

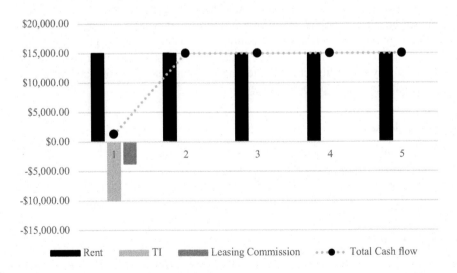

Figure 1.1 Cash flows associated with Example 1.8

Effective rent

In the last example, we saw that lease present value increases with the term of the lease. However, a landlord wouldn't necessarily wish to offer a tenant low rent for a very long term just to increase the lease's present value. This is because it may be possible to do better by having a series of shorter leases signed by tenants paying higher rent. For this reason, we would ideally like to have a measure of lease value that reflects the value of a lease to the landlord *each year*. That is what motivates the calculation of effective rent.

Effective rent is defined as the annual, constant cash flow that would deliver the same lease present value as any given lease. That is, calculating effective rent converts a lease's promised payments into an annual figure. Let's calculate effective rents for each of the 3 example leases described previously. In Example 1.8, we calculated that 5 payments of $15,000 were partially offset by an initial $10,000 TI and a $3750 leasing commission. Therefore, the property owner would only receive $1250 in the first year, followed by $15,000 in each of the following 4 years, which as calculated delivers a net present value of $44361.80. The effective rent of this lease is the level of payment that, if received each year during the lease's term, would generate the identical present value. The Excel command PMT (10%, 5, 44361.80) calculates that, at a 10% discount rate, a payment of $11,702.53 in each of the 5 years delivers the same present value. The comparison of the actual cash flows and the effective rent that generates the same present value is illustrated in Figure 1.2. Analogous calculations yield an effective rent of $16,402.76 for the second 5-year lease, but only $8332.39 for the 50-year lease. This lower effective rent better reflects the value of the long-term lease since the rental rate is noticeably lower than the others.

Note that similar effective rent calculations can also be made even if there are rental rate adjustments over the life of the lease.

Example 1.11: Suppose the lease from Example 1.8 called for annual rent increases of 1% per year. You can calculate that its net present value rises to $45,332.04 and its effective rent increases to $11,958.48.

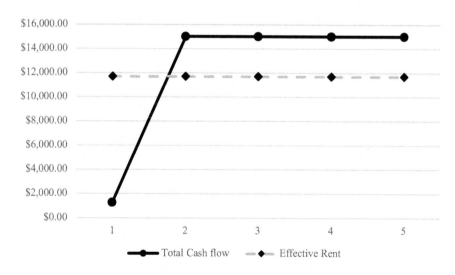

Figure 1.2 Actual cash flows and effective rent associated with Example 1.8

Other factors in lease choice

Given that effective rent converts the financial benefits and costs of a lease into annual amounts, you may be tempted to believe that as a property investor, you should always prefer tenants to sign leases that maximize the effective rent you collect. In this section, we discuss reasons why this isn't entirely true. These include:

1. Leasing costs
2. Tenant risk
3. Leasing risk
4. Portfolio considerations
5. Synergies

Leasing costs influence the cash flows that landlords receive. Tenant and leasing risks, portfolio considerations, and synergies potentially influence both the level of cash flows and how these cash flows are valued.

Leasing costs

Sometimes, people will calculate effective rent simply as the annualized value of rent without consideration of leasing expenses. Because of this, they will then argue that maximizing effective rent should not be the only goal of an investor because it may be worthwhile to sacrifice effective rent in exchange for a longer term so that leasing costs are reduced. In our formulation, however, leasing costs have been explicitly included in our effective rent calculations. Therefore, reduction of explicit leasing costs is not a reason to avoid using effective rent as a guide to lease choice.

Having said that, there is at least one important cost not captured by our effective rent calculation that influences a landlord's choice of lease. That is **vacancy**. When a tenant's lease expires and the lease is not renewed, there is typically a period during which the given space in the property is vacant and not generating revenue for the property owner. Effective rent comparisons across leases implicitly assume that the effective rent is collected each and every year without interruption. If shorter leases imply higher vacancy over time, it may be worth sacrificing effective rent for a lease with a longer term to minimize periods during which no rent is collected.

Tenant risk

Our calculation of effective rent was dependent on an assumption regarding the appropriate discount rate to use for discounting lease cash flows. Although our discussion of how to choose an appropriate discount rate will wait until Chapter 2, for now it is enough to realize that although using a common discount rate to calculate effective rents for different potential leases to the *same* tenant may be reasonable, it would not be reasonable to use the same discount rate when comparing leases being offered to *different* tenants if those 2 tenants presented different risks to the property owner. The appropriate discount rate used to calculate effective rent of a given lease reflects **within-lease risk**.

Leasing risks

Another consideration is the fact that vacancy presents additional risk to the property investor. Consider 2 leases promising identical effective rents, but one lease is for 5 years and the other is for 10 years. Not only does the 5-year lease present a greater likelihood of incurring vacancy, but the vacancy itself creates additional risk. Even if the 2 leases have identical within-lease risk, the shorter lease presents the property owner with higher **across-lease risk**, which is the risk that the owner will have difficulty filling the space after the initial lease term. All else equal, shorter-term leases generate higher across-lease risk, which reduces the value of the lease to the landlord. Thus, property owners may be willing to sacrifice effective rent for a longer term to reduce the risk of the property's cash flows after the leases expire.

> Example 1.12: Suppose a property investor is debating offering a prospective tenant a lease with either a 5-year or a 10-year term. In both cases, rent, TIs, and leasing commissions are chosen to deliver an identical effective rent of $30,000 per year. Suppose that if a 5-year lease is offered, the property investor expects to be able to either renew the existing tenant or sign a new tenant for the same effective rent of $30,000 per year for an additional 5 years. Note that the expected effective rent over the subsequent 10 years is the same in both instances. However, the risk faced by the landlord is not. Even assuming all tenants have the same within-lease risk, the cash flow received during the second 5-year period is safer as part of a 10-year lease as opposed to being from a future 5-year lease. The 5-year lease presents the landlord with across-lease risk. Suppose that within-lease risk warrants an 8% discount rate and across-lease risk warrants a 12% discount rate. The present value of a 10-year lease can be calculated as $201,302.44. However, the present value of 2 consecutive 5-year leases is only $187,748.43. This latter figure is calculated as the NPV of the first 5 years of the lease discounted at 8% plus the NPV (as of time 5) of the second 5 years (which is the same as the NPV of the first 5 years) discounted at 8%, then discounted back to time 0 at the higher 12% discount rate. These calculations are illustrated in Figure 1.3. Note that since the present value of two 5-year leases is less than that of a 10-year lease with identical effective rent, a property owner may be willing to sacrifice effective rent in exchange for a longer lease.

Portfolio considerations

As highlighted, vacancy is costly to a property investor and presents both direct costs (lost rent) and indirect costs (higher across-lease risk). Property investors often seek to avoid concentrations of vacancy as a way to manage these vacancy costs. Therefore, an owner of a multi-tenant property might be willing to sacrifice effective rent to ensure that multiple leases are not expiring at the same point in time.

Synergies

Suppose you own a small office building. You are leasing most of your space to a small law firm, but there is some vacancy. Would it matter to you if you leased the remaining space to a software developer or to a financial planning firm? You might be thinking that these 2 different tenants might have different risks, but to abstract from that, suppose that the correctly calculated effective rents for the 2 tenants are identical and that both tenants are interested in the same lease term. Under those circumstances, it may not matter which you choose because the cash flows you receive from the law firm are independent of the cash flows you may receive from either the software developer or the financial planners.

10 *Leasing*

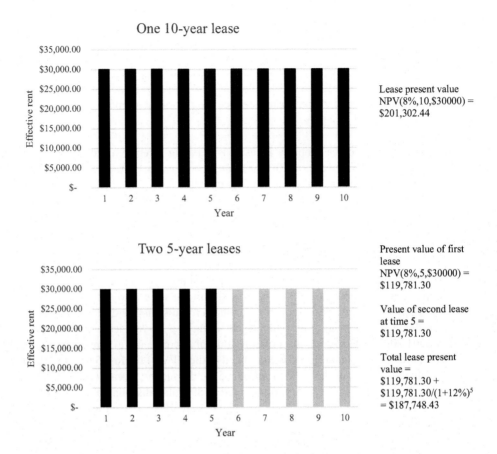

Figure 1.3 Lease present value with within- and across-lease risk

    Contrast that circumstance with one where you own a small retail shopping center. Your current tenants all cater to wealthy customers. For example, you might be leasing to a luxury jeweler, a designer handbag store, and a top fashion designer. You have 1 last space in your center and again you are considering 2 potential tenants who will deliver the same effective rent and want the same lease term. However, one tenant will use the space as a mini-showroom for luxury electric cars, while the other sells a variety of products, all of which sell for $1. Now do you have a preference? In this case, it is likely that you do. You likely prefer the luxury car showroom because the people who are shopping for luxury cars may likely shop at your existing tenants too. There is little chance that customers of the discount shop will also be looking at luxury goods. In a retail setting, synergies between different tenants can improve the performance of *all* tenants, which likely reduces the risk faced by the property owner and may explicitly increase the landlord's cash flow if the leases contain percentage rent agreements. It is the possibility of these synergies that motivate percentage rent agreements in a retail setting in the first place! Synergies between tenants at a shopping center can be valuable by reducing tenant turnover and the resulting vacancy and by potentially reducing the risk of the tenants as well. Thus, a property investor may be willing to sacrifice effective rent to sign tenants that increase synergies with existing tenants.

Leasing decisions in practice

In this case, you are put in the shoes of Benedict Clarke, owner of a 60,000-square-foot retail shopping center outside of Orlando, Florida. A recent bankruptcy has caused Clarke to lose his anchor tenant, and he faces significant inline vacancy as well. Clarke's leasing decisions will determine the fate of the shopping center moving forward. After reading and analyzing the case, you will be able to:

1 Calculate the NPV of a lease and convert it into an effective rent.
2 Understand the limitations of using effective rents to drive lease decision making.
3 Qualitatively and quantitatively analyze a set of prospective tenants with the goal of making optimal leasing decisions in a retail real estate context.

Tulaberry Plaza: leasing decisions in commercial real estate

"This isn't as easy as I thought it would be," Benedict Clarke murmured to himself. It had been only 3 years since Clarke had purchased a thriving retail shopping center outside of Orlando, Florida. Now, in January 2019, his anchor tenant had declared bankruptcy and vacancies in his inline spaces had been more difficult to fill than he had anticipated given the rise in online shopping.

The property

Tulaberry Plaza was a 59,100-square-foot gross leasable area (GLA) neighborhood shopping center located in the Southeast Orlando submarket. The property contained 2 buildings, with 8650 and 50,450 square feet of leasable space, respectively. Located nearby, but not part of the property, were 2 freestanding buildings currently occupied by banks. The property was located across the street from a Publix grocery store.

Clarke's initial years of ownership had been rather uneventful, but the property's anchor tenant, a nationwide electronics retailer, had recently vacated its space and ceased paying rent. Looking over the property's site plan (Figure 1.4) and rent roll (Table 1.2), Clarke realized that he would have to address the shopping center's vacancies. He also considered that the leases of some of his existing tenants would soon expire. Wishing for an existing tenant to vacate might have seemed counterintuitive, given the changes that were undoubtedly facing the shopping center. Nevertheless, Clarke wondered whether upcoming lease expirations were an opportunity to change the overall tenant mix. Of course, his experience suggested that every vacated space typically took 6 months to fill and required potentially significant tenant-specific improvement expenditures (TIs).

Leasing opportunities and costs

With significant vacancy and lease renewals, Clarke realized that leasing commissions (LCs) and TIs were major costs to be considered in the operation of Tulaberry Plaza over the near term. His leasing broker, Alice Cho, charged a commission of 3% of total lease revenue (total rental payments less TIs) for new tenants, and 1.5% for renewing existing tenants. These commissions were due immediately upon the signing of a new lease.

TIs varied widely according to the type of tenant, so Clarke carefully reviewed a list of prospective anchor tenants that Cho had put together for him that outlined each tenant's

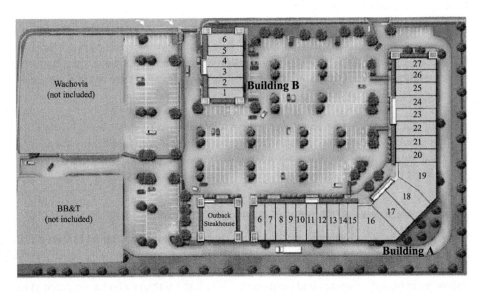

Figure 1.4 Site plan of Tulaberry Plaza

Table 1.2 Rent roll of Tulaberry Plaza

Unit	Tenant	Industry	Size (sq.ft.)	Lease End Date	Current Rent per Sq.Ft. ($)	Current Annual Rent ($)	Annual Rent Adj. (%)
B-1	Quizno's	Submarine sandwich/ salad restaurant	1400	12/31/25	32.00	44,800	2
B-2	*Vacant*		1400				
B-3	Papa John's	Pizza take-out/ delivery	1400	12/31/21	30.00	42,000	2
B-4	Jamba Juice	Drink-oriented restaurant	1400	12/31/20	30.00	42,000	2
B-5	Great Clips	Unisex hair salon	1260	12/31/19	30.00	37,800	2
B-6	Nature's Table Café	Café-style restaurant	1790	12/31/25	30.00	53,700	2
A-1–5	Outback Steakhouse	Restaurant	6401	12/31/32	10.15	65,000	2
A-6	Pike & Co.	Title and mortgage company	1400	12/31/22	28.00	39,200	2
A-7	Polo Cleaners	Dry cleaners	1400	12/31/20	32.00	44,800	2
A-8	Pak Mail	Shipping and packaging	1400	12/31/21	28.00	39,200	2
A-9	Deep Nine Salon & Spa	Hair salon and day spa	1400	12/31/23	28.00	39,200	2
A-10	*Vacant*		1386				
A-11	Total Family Care	Family medical practice	1337	12/31/26	28.00	37,436	2
A-12	Family Chiropractic	Chiropractic office	1400	12/31/19	28.00	39,200	2

(*Continued*)

Unit	Tenant	Industry	Size (sq.ft.)	Lease End Date	Current Rent per Sq.Ft. ($)	Current Annual Rent ($)	Annual Rent Adj. (%)
A-13	Lakeside Pediatrics	Pediatric medical practice	1387	12/31/27	28.00	38,836	2
A-14	*Vacant*		1387				
A-15	*Vacant*		1388				
A-16–19	*Vacant*		14228				
A-20	Chinese/Thai	Chinese and Thai restaurant	1984	12/31/26	28.00	55,552	2
A-21	Florida Fever	Clothing and home furnishings	1984	12/31/21	28.00	55,552	2
A-22	Animal Hospital	Veterinary medical facility	1984	12/31/25	28.00	55,552	2
A-23	Home Meal Replacements	Catering	1984	12/31/27	28.00	55,552	2
A-24	Insurance Office	Insurance office	2000	12/31/26	28.00	56,000	2
A-25	Vision Center	Optometry, optical devices	2000	12/31/19	26.00	52,000	2
A-26	Dentistry Office	General dentistry office	2000	12/31/27	28.00	56,000	2
A-27	Sushi Style	Asian sushi restaurant	2000	12/31/29	28.00	56,000	2

specific lease terms, including their TI needs (Table 1.3). Clarke noticed immediately that the lease terms offered by the potential anchor tenants differed not only in terms of required TIs but also in terms of base rent and term. Cho had also provided similar information for potential inline tenants (Table 1.4).

Clarke understood from Cho that the most important decision he needed to make was how to fill the anchor space in the shopping center. An anchor tenant tends to drive traffic to other stores in a plaza and, as a result, is typically offered lower rent on a per-square-foot basis. An anchor's ability to drive traffic to inline stores depends crucially on the similarity of the anchor's typical customer and the demographics of the local population.[2] Thus, Clarke had to give serious thought to which tenants would be best for Tulaberry Plaza.

The decision

Clarke understood that he needed to act quickly. Because his tenants had NNN leases, Clarke didn't personally incur any operating expenses. However, he was responsible for any capital expenditures on the property, which historically averaged around $85,000 per year. He also had annual mortgage payments on the property of $750,000 per year. He certainly didn't want to think about what might happen if he didn't make every mortgage payment on time. As he sat down at his desk to start crunching some numbers, the words "Skilled financial analysts can make a spreadsheet to justify anything – so think carefully about your assumptions," echoed through his head. If only he could recall where he had first heard them. . . .

Table 1.3 Potential anchor tenants

Tenant	Walgreens	Harbor Freight	PetSmart	Planet Fitness
Tenant type	Anchor	Anchor	Anchor	Anchor
Industry	Pharmacy	Hardware store	Pet store	Fitness
Creditworthiness	S&P: BBB	S&P: BB-	S&P: CCC	Franchisee
Projected sales (per SF)	$572	$308	$289	$78
Tenant improvements requirement (per SF)	$35.00	$27.00	$28.00	$44.00
Lease type	NNN	NNN	NNN	NNN
Term	25 years, with 5 additional 5-year options to extend at $25.00	15 years	10 years	20 years
Base rent amount (per SF)	$15.00	$18.75	$18.50	$21.00
Rent escalations	0%	2% annual increase	2% annual increase	10% increase after every 5 years
Target demographics	Male and female, all ages, all incomes	Male, ages 25–55, wide income range	Female, ages mid-20s and 50s	Male and female, ages 18–45, low- to middle-income
Location preference	Near intersection/ drive-through	Visible, inline, or self-standing	Anchor	Anchor
Parking requirement	Medium	Medium	Low	High (100 spaces)
Size requirement	10,800–15,000	12,000–15,000	18,000–27,500	12,000–20,000

Table 1.4 Inline tenant possibilities

Tenant	Starbucks	The Vitamin Store	Ben's Tux Rental	Family Dollar
Tenant type	Restaurant	Inline	Inline	Inline
Industry	Coffee/food	Health and wellbeing	Formalwear rental	Value retail
Creditworthiness	S&P: BBB+	N/A	N/A	S&P: BBB-
Projected sales	N/A	N/A	N/A	N/A
Tenant improvements requirement (per SF)	$39.00	$25.00	$20.00	$22.00
Lease type	NNN	NNN	NNN	NNN
Term	10 years	10 years	5 years	10 years
Base rent amount (per SF)	$33.75	$25.50	$24.00	$24.50
Rent escalations	10% every 5 years	10% every 5 years	3%	10% increase every 5 years
Target demographics	Male and female, ages 25–40, mid-income ($75k or more)	Male and female, educated, health enthusiasts	Male, ages 18–30	Male and female, low- and middle-income, wide age range
Location preference	Urban or suburban, on commuting side of traffic patterns	Highly visible	Cheap	Food store anchored, neighborhood shopping center
Parking requirement	Medium	Low	Low	Medium (25+ spaces)
Size requirement	1,700–2,700	3,000	1,200–1,500	7,000–10,000

(*Continued*)

Table 1.4 (Continued)

Tenant	Salons	Shoe Village	Local Furniture Boutique	Five Guys
Tenant type	Inline	Inline	Inline	Restaurant
Industry	Hair salon and day spa	Footwear	Home furnishings	Food
Creditworthiness	3-month security deposit	N/A	3-month security deposit	Franchisee
Projected sales	N/A	N/A	N/A	N/A
Tenant improvements requirement (per SF)	$22.00	$19.00	$20.00	$48.00
Lease type	NNN	NNN	NNN	NNN
Term	5 years	15 years	5 years	5 years
Base rent amount (per SF)	$25.50	$23.50	$25.50	$33.75
Rent escalations	3%	10% increase every 5 years	3%	3%
Target demographics	Female, middle-income	Female, ages 18–44, household income less than 75k	Male and female, ages 18–45, low- to middle-income	Young adults
Location preference	None	None	None	Corner or end cap ideal, but will consider inline
Parking requirement	Low	Low	Low	Medium (35 spaces)
Size requirement	1,200–1,500	6,000–15,000	1,200–1,400	2,000–3,000

(*Continued*)

Table 1.4 (Continued)

Tenant	Salons	Shoe Village	Local Furniture Boutique	Five Guys
Tenant type	Inline	Inline	Inline	Inline
Industry	Gaming	Travel agency	Clothing retail	Wireless service
Creditworthiness	S&P: BB+	3-month security deposit	3-month security deposit	S&P: BBB+
Projected sales	N/A	N/A	N/A	N/A
Tenant improvements requirement (per SF)	$35.00	$25.00	$24.00	$37.50
Lease type	NNN	NNN	NNN	NNN
Term	5 years	5 years	10 years	10 years
Base rent amount (per SF)	$30.00	$25.50	$25.50	$40.00
Rent escalations	3%	3% per year	3% per year	0%
Target demographics	Male, ages 15–30	Male and female, ages 35–55	Male and female, ages 12–24	Male and female, ages 25 and older
Location preference	None	None	None	Outlier/inline/endcap
Parking requirement	Medium	Low	Medium	Low
Size requirement	1,200–1,400	1,200–1,400	2,800–3,600	2,600–4,600

Notes

1 Note that we are assuming that the TIs and LCs are paid in Year 1, at the same time that the rent is first collected. This is a simplification that we will use to ease our calculations. In practice, TIs and LCs would be paid before the entire first year's rent was collected.
2 There were currently just over 5300 people (2200 households) living within 3 miles of the shopping center, with a median age of 43, household income of $104,000, and home price of $403,000.

2 Property valuation

Introduction

"What's it worth?" That is the most fundamental question an investor faces when considering the purchase of an investment property. In this chapter, we argue that the value of an existing property can be estimated by calculating the present value of the expected cash flows it delivers to its owner. For a student of finance, this is not a new idea. To apply this idea to investment property requires answering the following questions:

1. How much cash flow to the investor will the property generate?
2. At what rate should those cash flows be discounted back to today?

Cash flow from investment property

What are the sources of cash flow from an investment property? If you ask an investor that owns a small apartment building this question, the answer that comes immediately to mind is rent. That is, investment property is valuable because the owner of the property collects rent from its tenants. What is important to consider is that although rent may be among the most important drivers of property value, it is only one of a number of sources of property income. The cash flows associated with the ownership of an investment property must also include a number of important expenses.

Another important fact that influences the value of an investment property is that property lasts a long time. This has 2 key implications. The first implication of the longevity of investment property is that a property's current value will be related to its ability to generate cash flow into the distant future. Thus, to estimate the value of an investment property today will require forecasting cash flows many years into the future. This will involve creating what is known in the industry as a cash flow **pro forma**, which is a projected financial statement that outlines the expected cash flows that will accrue to the owners of any given investment property. The second implication of the long-lived nature of investment property is that the property will typically outlive its current owner. That is, a typical real estate investor not only buys a given property, but also typically sells the property in the future. In our discussion of property valuation, it is useful to distinguish between cash flows arriving during the **holding period** – the time during which an investor owns the property – and **reversion cash flow** that arises from the ultimate sale of property.

Property valuation

Holding period cash flow

Gross potential rental income

When calculating operating income, it is typical to begin with a measure of gross potential rental income. **Gross potential rental income** combines rental income from 2 sources – space in the property that is currently under lease and space in the property that is not currently under lease. For example, suppose that you are considering the purchase of an industrial property that is currently configured into 2 spaces of 20,000 square feet each. One space is currently leased at $15. The other space is currently vacant. However, real estate brokers are telling you that the market for similar industrial space is currently leasing for $17 for a 10-year lease, the typical term for such space. The word *potential* in gross potential rental income implies that you consider both rental income that is promised under existing leases *plus* income you would receive if all vacant space were leased at market rates. The $17 mentioned by the broker is an estimate of what is known as **market rent**, the price that will be paid by a potential tenant to lease a particular type of space under current conditions prevailing in the local market. Market rent is influenced by the conditions present in both the local and broader economy, which in turn influences the supply and demand for space. Note, too, that "market" rent is influenced by factors that are specific to the space and not simply the market. For instance, higher floors in an office building might command a higher rent. Newer space might warrant higher rents than older space. Assume that all of these factors have been considered in the broker's estimate of $17.

> Example 2.1: Gross potential rental income for this particular property can then be calculated as the sum of the rent paid by the existing tenant ($15 × 20,000 = $300,000) and the rent that potentially could be collected from a new tenant ($17 × 20,000 = $340,000). This sums to $640,000.

Note that you can (and generally would) calculate gross potential rental income for future dates as well.

> Example 2.2: Suppose that industrial tenants in the property's market sign 10-year leases with flat rent. Further suppose your existing tenant had 2 remaining years on its lease and that you anticipated that market rent would be $18 after 2 years. Your gross potential rent would remain $640,000 for the first 2 years, but would then rise to $700,000 beginning in Year 3. This is because you are assuming that the space currently being leased at $15 will be re-leased at $18, bringing the potential rent for that space to $360,000 ($18 × 20,000) each year beginning in Year 3. The second space, which you assume would lease at today's market rate of $17, will still be paying $340,000 per year. Keeping track of these calculations in a spreadsheet, our calculation of gross potential rental income would look like what is shown in Table 2.1.

Table 2.1 Gross potential rental revenue

	Year 1	Year 2	Year 3	Year 4	Year 5	Year 6
Gross potential rental income						
Space 1	$300,000	$300,000	$360,000	$360,000	$360,000	$360,000
Space 2	$340,000	$340,000	$340,000	$340,000	$340,000	$340,000
Total gross potential rental income	$640,000	$640,000	$700,000	$700,000	$700,000	$700,000

Deductions from potential rental income

In our calculation of gross potential rental income, we combined income from tenants with leases and income from vacant spaces. In other words, we calculated a number that represents the fullest potential of the property to generate rental income. In reality, an owner will not realize rental income equal to the property's fullest potential. Generally, we don't expect our investment property to be fully leased, we offer concessions on our rent to attract tenants, and we don't expect all of our tenants to pay us every dollar that they promised us in their lease agreement. Therefore, we need to make deductions for vacancy allowances, rental concessions, and credit reductions.

> Example 2.3: Suppose that it will take 9 months for us to find a tenant to lease Space 2 at the current market rate of $17. Further suppose that to entice our new tenant to sign its lease for Space 2, we offered the tenant its first 3 months of rent free. We anticipate that market conditions for leasing will be stronger in the coming years, so we do not anticipate needing to grant a concession to the tenant we will attract in Year 3. Finally, suppose that we understand our existing tenant's business is not doing well and therefore anticipate that the company will go out of business during Year 2 and skip its final 2 rent payments.

A **vacancy allowance** corrects for the fact that we will not be collecting rent from all of our available space at all times. In this example, we would need to deduct vacancy allowances of $255,000 in Year 1 for the 9 months that Space 2 is vacant ($17 × 20,000 × 9 months out of 12). Further suppose that the tenant in Space 1 does not renew its lease when it expires and that it will take 6 months to find a replacement tenant who will pay the $18 market rent. We would need to deduct a vacancy allowance of $180,000 in Year 3 for the 6 months that Space 1 is vacant ($18 × 20,000 × 6 months out of 12). Offering 3 months free rent is an example of a rent concession discussed in Chapter 1. This would cause further deduction from potential rental income equal to $85,000 ($17 × 20,000 × 3 months out of 12). Finally, we might also make a **credit reduction** for the possibility that our tenants do not ultimately pay us all of the rent that was promised to us in their leases. We would therefore collect $50,000 less than promised from Space 1 in Year 2 ($15 × 20,000 × 2 months out of 12). Of course, if our tenant fails to pay us rent as promised, we would have a legal claim against the assets of the tenant's company for the amount of 2 months' rent. The value of this claim depends on the value of the tenant's business and the claims of its other creditors. This example abstracts from this issue.

Forecasting vacancy and credit deductions can be done in a number of ways. An investor might consider looking at historical deductions for the given property or the level of vacancy in the market for similar properties. For properties with few tenants, such as our example industrial property, it is typical to estimate vacancy and credit deductions on a space-by-space basis by estimating renewal probabilities and the time between tenants during which space would be vacant. Finally, local, regional, and national business cycles are likely to affect not only market rents but also vacancy and credit deductions.

Effective net rental income

Effective net rental income is the rental income expected to be received by the property owner. This can be calculated directly as the gross potential rental income less deductions for vacancy, rent concessions, and credit. Subtracting our 3 types of deductions from gross

22 Property valuation

Table 2.2 Effective net rental income

	Year 1	Year 2	Year 3	Year 4	Year 5	Year 6
Gross potential rental income						
Space 1	$300,000	$300,000	$360,000	$360,000	$360,000	$360,000
Space 2	$340,000	$340,000	$340,000	$340,000	$340,000	$340,000
Total gross potential rental income	$640,000	$640,000	$700,000	$700,000	$700,000	$700,000
Vacancy allowance						
Space 1	$0	$0	$180,000	$0	$0	$0
Space 2	$255,000	$0	$0	$0	$0	$0
Rental concessions						
Space 1	$0	$0	$0	$0	$0	$0
Space 2	$85,000	$0	$0	$0	$0	$0
Credit reductions						
Space 1	$0	$50,000	$0	$0	$0	$0
Space 2	$0	$0	$0	$0	$0	$0
Effective net rental revenue	$300,000	$590,000	$520,000	$700,000	$700,000	$700,000

potential rental revenue, we can calculate effective net rental revenue. Our spreadsheet now looks like what is shown in Table 2.2.

Other income

Investment property can generate income from additional sources other than rent from its tenants. For instance, landlords may collect additional income from parking, food concessions, and a variety of other sources. Suppose this industrial property has allowed a telecommunications company to place a cell tower on its roof. In exchange, the telecommunications company pays the landlord $2000 each month. As the property owner, we would want to account for the receipt of an additional $24,000 in annual revenue.

Another source of other income would be expense reimbursements from tenants. In this example, let us assume that our tenants have gross leases. Thus, there will be no reimbursement of expenses by the tenants to the landlord. If there were, the magnitude of the required reimbursements could only be estimated after considering the expenses associated with the operation of the property.

Operating expenses

There are a number of expenses associated with the ownership and operation of an investment property. The magnitude of these costs will vary across markets and property types. It is useful to distinguish between operating expenses in 2 dimensions. The first is whether or not the expense is generally reimbursable by the tenant. As described in Chapter 1, the specific expenses subject to reimbursement from the tenant will be specified in each tenant's lease.

A second dimension of operating expenses is whether the expense is fixed or variable. Fixed expenses are those that do not vary with the level of building occupancy. For instance, real estate taxes, property insurance, and building security may be considered to be fixed. Other expenses, such as maintenance and utilities, are variable. These expenses tend to fall as building occupancy falls.

In estimating total operating expenses, we would enumerate each expense and estimate how the expense might evolve over time, both because occupancy is forecasted to change and also because many expenses tend to slowly increase with inflation. In our example industrial property with 2 rentable spaces, it is reasonable to assume that expenses in Year 1 will be notably lower than in the following years since the building is expected to have significant vacancy. Likewise, you expect expenses in Year 3 to be lower than those in Years 2 and 4, also because of expected vacancy.

Net operating income (NOI)

Net operating income is a property metric that is defined as effective net rental income plus other income less operating expenses. It represents the operating cash flow expected to be earned by the property irrespective of who owns it, how it is financed, or the owner's personal **income tax** consequences. Consolidating figures across our 2 tenants and forecasting operating expenses, NOI for our example can be calculated as described in Table 2.3.

Capital expenditures

Building improvements are capital expenditures undertaken to significantly improve the quality of the property. One way to view building improvements is that they are the necessary expenses a property owner must undertake to offset the economic depreciation of the property. For example, a property owner will need to periodically replace the property's roof or repave the property's parking lot.

Another feature of building improvements is that an owner typically has flexibility with respect to when such capital expenditures are undertaken. For example, the parking lot of our industrial complex may be in disrepair. As a property owner, you may believe that it should be replaced in Year 3, but another owner may believe it can wait until Year 4. The discretionary nature of building improvements leads some property owners to avoid specifying when certain large expenditures will be completed. Instead, the owner may set aside some money

Table 2.3 Net operating income

	Year 1	Year 2	Year 3	Year 4	Year 5	Year 6
Total gross potential rental income	$640,000.00	$640,000.00	$700,000.00	$700,000.00	$700,000.00	$700,000.00
Vacancy allowance	$255,000.00	$0.00	$180,000.00	$0.00	$0.00	$0.00
Rental concessions	$85,000.00	$0.00	$0.00	$0.00	$0.00	$0.00
Credit reductions	$0.00	$50,000.00	$0.00	$0.00	$0.00	$0.00
Effective net rental revenue	$300,000.00	$590,000.00	$520,000.00	$700,000.00	$700,000.00	$700,000.00
Other income	$24,000.00	$24,000.00	$24,000.00	$24,000.00	$24,000.00	$24,000.00
Effective gross income	$324,000.00	$614,000.00	$544,000.00	$724,000.00	$724,000.00	$724,000.00
Operating expenses	$139,320.00	$208,760.00	$206,720.00	$242,540.00	$246,178.10	$249,870.77
Net operating income (NOI)	$184,680.00	$405,240.00	$337,280.00	$481,460.00	$477,821.90	$474,129.23

24 *Property valuation*

each year in a **capital reserve account**. Money in such an account is usually targeted to the average amount of capital expenditures related to building improvements that happen in a given year.

Suppose in our industrial building example, a careful inspection of the property identifies that the roof is certain to need replacement in Year 3 at an estimated expense of $80,000.

Leasing costs are additional capital expenditures undertaken to attract or maintain tenants that are incurred when leases are signed. As we discussed in Chapter 1, property owners pay tenant improvement allowances (TIs) to upgrade or repurpose space in the property for the benefit of the tenant. Property owners negotiate TIs with prospective tenants, with TIs for new tenants looking to reconfigure space typically higher than TIs offered to existing tenants looking to update and modernize their space. For our industrial property, suppose we offer TIs of $60,000 to attract tenants in both Year 1 and Year 3.

Another leasing cost is **leasing commissions (LCs)**, which are fees paid to leasing brokers to compensate them for marketing the property, providing property tours, and identifying potential tenants to the property owner. The need to incur such expenses will be tied to the leasing needs of the property, which in turn is associated with the expiration of existing leases. In our industrial property example, suppose our leasing broker charges a commission of 5% of total lease revenue in the year that a new lease is signed. This costs the property owner $165,750 (5% × [$340,000 × 10 – $85,000]) in Year 1 and $180,000 (5% × 360,000 × 10) in Year 3.

Holding period cash flow

Holding period cash flow is the total amount of cash flow generated by the property during the investor's ownership. This is calculated as net operating income less capital expenditures. Adding our capital expenditures to our spreadsheet, we can calculate the property owner's holding period cash flow as shown in Table 2.4.

In this example, holding period cash flow is negative in Year 1. As will be explained in the next section, we will assume that the property is being sold at the end of Year 5, and for that reason holding period cash flow is not calculated beyond that point.

Reversion cash flow

In the previous section, we calculated the cash flows that an investor in an industrial property would expect to receive during the first 5 years of ownership. If the investor plans to own the property for additional years, then the calculations illustrated above could be continued. Eventually, an investor will sell the property. Suppose that the investor expects to sell the property at the end of Year 5. The sales price of the property less the commission paid to any

Table 2.4 Holding period cash flow

	Year 1	Year 2	Year 3	Year 4	Year 5
Net operating income (NOI)	$184,680	$405,240	$337,280	$481,460	$477,822
Building improvements	$0	$0	$80,000	$0	$0
TIs	$60,000	$0	$60,000	$0	$0
Leasing commissions	$165,750	$0	$180,000	$0	$0
Holding period cash flow	–$41,070	$405,240	$17,280	$481,460	$477,822

sales broker will be the reversion cash flow, defined as the cash flow received net of fees to the investor arising from the sale of the property.

To estimate reversion cash flow, we need to answer the question, "How much will the property sell for at the end of Year 5?" You might be confused by this question, since we have been discussing property level cash flow to answer the question, "How much is the property worth today?" Now it seems that the answer to how much the property is worth today is dependent upon how much the property is worth in 5 years. Let's set aside this circularity for a moment and introduce a convenient shorthand way to estimate a property's value that will prove useful.

Let us define an investment property's **cap rate** as the property's year-ahead net operating income divided by its current price. If we define the cap rate of a property at time t as c_t and the price of the property at time t as P_t, then the definition of a cap rate can be written:

$$c_t = \frac{NOI_{t+1}}{P_t}$$

Note that the cap rate is a measure of the income-producing capability of an investment property expressed as a percentage of property value. In that sense, it is similar to a dividend yield for a stock in that both express measures of the income generated by an asset in the coming year as a percentage of its current price. Note one important difference between a cap rate and a dividend yield is that a cap rate is calculated based on NOI, which is an income measure calculated before building improvements and leasing costs.

In our example, we were looking for the value of our industrial property at the end of Year 5. We can invert the definition of a cap rate to solve for the value of the property in Year 5 by dividing its Year 6 net operating income by the building's appropriate cap rate at the end of Year 5. Note that this relationship explains why it is common to estimate NOI for 1 year longer than the presumed holding period when valuing an investment property. The cap rate used to estimate the sale price of an investment property at reversion is known as the **going-out cap rate** or **exit cap rate**. We will return shortly to the very important questions, "From where does an investor get cap rates?" and "How should an investor use cap rates?" For now, let's assume the investor knows that an appropriate exit cap rate for this industrial property is 8% and that brokers to the sales transaction will keep 3% of the sales price as commissions. Assuming a 1.5% increase in operating expenses after Year 5, the spreadsheet forecast for cash flows can be expanded as shown in Table 2.5. **Total property cash flow** is calculated as the sum of reversion cash flow (sales price less sales commissions) and holding period cash flow.

Table 2.5 Total property cash flow

	Year 1	Year 2	Year 3	Year 4	Year 5	Year 6
Net operating income (NOI)	$184,680	$405,240	$337,280	$481,460	$477,822	$474,129
Building improvements	$0	$0	$80,000	$0	$0	
TIs	$60,000	$0	$60,000	$0	$0	
Leasing commissions	$165,750	$0	$180,000	$0	$0	
Holding period cash flow	–$41,070	$405,240	$17,280	$481,460	$477,822	
Reversion cash flow						
Sales price					$5,926,615	
Sales commission					$177,798	
Total property cash flow	–$41,070	$405,240	$17,280	$481,460	$6,226,639	

Property valuation

Table 2.6 The property pro forma

	Year 1	Year 2	Year 3	Year 4	Year 5	Year 6
Gross potential rental income						
Space 1	$300,000	$300,000	$360,000	$360,000	$360,000	$360,000
Space 2	$340,000	$340,000	$340,000	$340,000	$340,000	$340,000
Total gross potential rental income	$640,000	$640,000	$700,000	$700,000	$700,000	$700,000
Vacancy allowance						
Space 1	$0	$0	$180,000	$0	$0	$0
Space 2	$255,000	$0	$0	$0	$0	$0
Rental concessions						
Space 1	$0	$0	$0	$0	$0	$0
Space 2	$85,000	$0	$0	$0	$0	$0
Credit reductions						
Space 1	$0	$50,000	$0	$0	$0	$0
Space 2	$0	$0	$0	$0	$0	$0
Effective net rental revenue	$300,000	$590,000	$520,000	$700,000	$700,000	$700,000
Other income	$24,000	$24,000	$24,000	$24,000	$24,000	$24,000
Effective gross income	$324,000	$614,000	$544,000	$724,000	$724,000	$724,000
Operating expenses	$139,320	$208,760	$206,720	$242,540	$246,178	$249,871
Net operating income (NOI)	$184,680	$405,240	$337,280	$481,460	$477,822	$474,129
Building improvements	$0	$0	$80,000	$0	$0	
TIs	$60,000	$0	$60,000	$0	$0	
Leasing commissions	$165,750	$0	$180,000	$0	$0	
Holding period cash flow	−$41,070	$405,240	$17,280	$481,460	$477,822	
Reversion cash flow						
Sales price					$5,926,615	
Sales commission					$177,798	
Total property cash flow	−$41,070	$405,240	$17,280	$481,460	$6,226,639	

The property pro forma

Although we have constructed the property's cash flows in steps, generally speaking all of the forecasting would be combined in a single pro forma, which is a model of all relevant cash flows on a going forward basis. That is, one would typically see a model with all of the cash flows associated with the investment property in a single place, such as what is illustrated in Table 2.6.

Investment property valuation

Discounted cash flow

Our earlier discussion outlined how one could forecast the cash flows accruing to a property investor. To complete a property valuation using the discounted cash flow technique requires determining the appropriate discount rate for those cash flows. This section describes a variety of approaches to approximate the appropriate discount rate to use for a property valuation. Some of these approaches cannot be applied to an individual property, but rather are generally used to estimate discount rates using data from a portfolio of properties broadly similar to the property being valued. With that in mind, property-specific adjustments will likely be required. Further, it is generally a good idea to approach the selection of a discount rate in a variety of ways to gain confidence in your ultimate choice.

The CAPM

The workhorse model of corporate finance is the **capital asset pricing model (CAPM)**, which argues that the risk premium on a risky asset should be proportional to its **beta**, a measure of how the asset adds risk to an otherwise diversified portfolio. The problem with using the CAPM to estimate risk premiums in investment property is that it is very difficult to calculate returns on an individual property or even a portfolio of assets on a frequent basis. Unlike stocks that trade nearly continuously, an investment property may trade only once every 5 or 10 years. It may, however, be possible to estimate betas for portfolios of properties that are similar to the subject property. Likewise, one might learn something about the risk of private real estate investments from the risks associated with investing in public real estate companies. These public companies have actively traded stock, from which an equity (and therefore asset) beta can be estimated.

Although it is difficult in practice to implement the CAPM on an individual property basis, the intuition in the model should always be your guide to understanding the use of discount rates in practice.

> Example 2.4: Historical data indicate that office property located in central business districts (CBDs) have historically returned less than office property in suburban locations. If we believe that CBD neighborhoods are more desirable for office tenants, all else equal, then such locations will have lower sensitivity to business cycles, and therefore a lower beta.
>
> Example 2.5: The relationship between discount rates and releasing risk can be explained using the intuition of the CAPM. The risk of finding new tenants upon the expiration of an existing lease is a risk that is not idiosyncratic. In particular, it should be positively correlated with the state of the real estate market, which in itself is systematically related to the economy in general. Therefore, all else equal, properties with greater lease turnover will require a higher discount rate.

Tenant risk

Generally speaking, the discount rate used in property valuation should reflect the underlying systematic risk of the property-level cash flows. Because most of the cash flow arises from rents received from tenants, the systematic risk of the tenants may be a useful guide to the appropriate discount rate. For example, if you are leasing 100% of your space to the US government, then it may be appropriate to use a lower discount rate than if you are leasing to local business owners. Large corporate tenants might warrant a discount rate somewhere in between.

Survey evidence

It is possible to estimate discount rates for investment properties simply by asking investors in a survey. Several national real estate firms regularly survey the investment community and publish the average stated return investors say they expect going forward.

Historical returns

Data on historical returns on real estate investments of various types is obtainable, and these can also be a useful guide to return expectations going forward. To the extent that

28 Property valuation

the systematic risks of a given property do not change significantly over time, historical average returns may be a useful indicator of average returns going forward. Unlike mutual fund managers that change portfolios from day to day, an office building is likely to remain an office building for a long time. Thus, historical returns can be a useful guide to the future because the building's risk, as measured by its beta, is likely to be relatively constant over time.

One caveat to this is that it may be useful to consider how the risk-free rate has changed over time to make adjustments to your discount rate estimate. That is, historical returns reflect the sum of past risk-free rates and risk premiums. That investment property has a constant beta implies a reasonably steady risk premium. If current risk-free rates are noticeably lower than they were during the period over which you have gathered return data, then an appropriate estimate of discount rates going forward would be lower than average historical returns.

Cap rates

Like historical returns, data on historical cap rates is often observable by property type and location. These cap rates can be used to guide the choice of discount rates. It is shown in this chapter's appendix that in the long-run, the discount rate r is related to the cap rate c by the expression $r = c(1 - \phi) + g^{NOI}$. In this equation, ϕ is the fraction of NOI spent on capital expenditures and g^{NOI} is the long-run growth rate of property NOI. Intuitively, the relationship says that the return on property comes from its cash flow yield (cap rate less capital expenditures) and its future growth.

Valuation

In our industrial property example, suppose that current survey evidence suggests that industrial property investors are seeking an 8% return going forward. We might adjust that to 8.5% because our property has greater than typical lease turnover in the near term. So, one estimate of the appropriate discount rate is 8.5%. Using another approach, suppose that the historical return on industrial property in the market is 8.75%, which we again might adjust to 9.25% because of our expected tenant turnover. Thus, a second estimate for the appropriate discount rate is 9.25%. Finally, suppose we observe that cap rates on industrial property in the area have recently averaged 7.5%. We also estimate that this property will require 20% of its NOI to be spent on capital expenditures and that we will be able to raise rent 25% every 10 years, or roughly 2.5% per year. Thus, a third estimate of an appropriate discount rate would be 8.5% = [7.5% × (1 − 20%) + 2.5%]. So, we have estimated discount rates of 8.5%, 9.25%, and 8.5%, and thus it may be reasonable to settle upon a choice of 8.75%. Applying an 8.75% discount rate to the total property cash flows derived earlier yields an estimated property value of $4,756,166.

Direct capitalization

Completing a discounted cash flow analysis of an investment property is a significant undertaking. Wouldn't it be great if there was a much simpler way? Someday you may encounter someone that seems to be able to estimate the value of an investment property in a matter of seconds. Here is the secret. We defined a property's cap rate to be equal to the year-ahead NOI divided by the property's current value. Inverting this relationship would imply that today's value of the investment property is simply its expected Year 1 NOI divided by its current cap rate. Thus, an estimate of property value can be derived without worrying about forecasting any cash flows beyond Year 1. The **direct capitalization method of property valuation** simply divides a measure of first-year NOI by an estimate of a current cap rate. The current cap rate is alternatively known as an **entry cap rate** or a **going-in cap rate** because it reflects the cap rate in the market at the time you become an investor in a given property.

Estimating the current cap rate of a building is generally an exercise in finding recent property transactions in the market for the property type being considered. Ideally, you would find properties that were also of similar size, quality, location, and other characteristics similar to the property you were trying to value. The comparable properties would also have similar lease maturities, rent escalators, tenant quality, etc. Thus, as a practical matter it will be impossible to find truly comparable properties.

Suppose we were trying to estimate the cap rate for a 14-story office building in a Northwest suburb of Chicago, IL. In that case, practitioners would typically take an average "market" cap rate, which, depending on market conditions, might be able to be calculated based on recent transactions of "office properties in Northwest suburban Chicago." This market average cap rate would then be adjusted based on how characteristics of the subject property differ from the characteristics of the set of properties used in calculating the market average. If the subject property were "better" than the typical building that had transacted in the market, then a practitioner would tend to lower the assumed cap rate. If the property were "worse," one would tend to raise the cap rate.

We can apply this valuation method to our example property. Suppose that in the *current* market for our particular industrial complex, other industrial buildings had traded for an average cap rate of 6.5%. However, the subject property has atypically high vacancy forecasted for the first year and so an investor might argue that "true" NOI for the property is closer to $400,000. Further the practitioner might note that the near-term lease expirations are higher than typical and as a result, they would use a higher cap rate of 7.5%. Combining these assumptions, a practitioner might quickly estimate the industrial building to be worth approximately $5.3 million ($400,000 ÷ 7.5%).

Reconciling the different approaches

The direct capitalization approach to property valuation assumes that a value can be accurately estimated from only 2 numbers – the first-year NOI and a cap rate. In particular, it assumes that property value is a multiple (equal to 1 ÷ c) of the first-year NOI. Assuming first-year NOI can be estimated accurately, uncertainty about property valuation will arise from uncertainty about the appropriate cap rate. If cap rates are derived from comparable properties (as is typically the case), anything that differs between the subject property and the comparable properties will necessitate an adjustment to the cap rate. In practice, this may be difficult to do.

If a discounted cash flow approach is applied correctly, then it will deliver an accurate property value. Under what conditions will the DCF valuation also be equal to a multiple of first-year NOI? It turns out that this will be true when ϕ, the fraction of NOI spent on capital expenditures, is constant and when NOI is growing at a constant rate g. Thus, using direct capitalization as a valuation technique essentially implies that the property is already in its "long-run" state of constant growth. This may be approximately true for fully occupied properties where you don't anticipate large deviations from constant NOI growth. For other properties, like our industrial complex example, this is very far from reality, and thus a DCF approach is to be preferred.

30 *Property valuation*

Using a market-based cap rate to estimate a property's worth is an exercise in determining the *price* at which the property is likely to transact. This is different from the DCF approach, which is seeking to determine the *value* of the investment property. One hopes that price is close to value, but experience suggests that this may not always be the case. If property values are affected by a bubble, this would tend to be reflected in abnormally low cap rates. Direct capitalization may accurately tell you a *price*, but it would not protect you from overpaying if prices later adjust downward to reflect their true value. Thus, one benefit of completing a DCF is to force you to consider the fundamental ability of a property to generate cash flow and to think realistically about the appropriate rate of return that those cash flows must deliver.

Finally, it is worth noting that the use of cap rates is commonplace among real estate professionals. For example, one might observe cap rates on office buildings in a market to be higher than cap rates on retail. This means that investors are paying more for NOI coming from a retail center than NOI from an office building. Why would they do this? Well, it could mean that investors believe retail cash flows are safer than office cash flows, or it could mean that retail cash flows are expected to grow more quickly than office cash flows. Another professional investor may comment that apartment cap rates have recently risen in a given market. Rising apartment cap rates imply that investors are paying less for a dollar of apartment NOI than they used to. Why? This too can be explained by changes to fundamentals. It could be that apartment cash flow is now viewed as riskier than it used to be or that investors have reduced their growth forecasts for apartment NOI. These examples illustrate that although you may hear people discuss cap rates as if they were a fundamental characteristic of an investment, it is important that a property investor understands that more fundamental features of a property investment, such as growth and risk, affect a property's value, which in turn affects the property's cap rate.

Final thoughts on investment property valuation

Choosing the horizon of the cash flow pro forma

For our example industrial building, our pro forma assumed a 5-year holding period. Would our valuation be any different if we adjusted our assumed holding period? If the valuation exercise has been done correctly, then the answer to this question is no. Imagine that we anticipate owning the industrial property for 20 years. We could have forecasted cash flows for 20 years, applied an exit cap to Year 21 NOI to estimate reversion value in Year 20, and applied a discount rate to value those cash flows as of today to estimate today's value of the property. Hopefully, you can see that valuing our property with a 5-year pro forma and valuing our property with a 20-year pro forma will give the same valuation today if our assumed reversion value in Year 5 in the 5-year model is equal to the present value of the cash flows modeled in Years 6–20 in the 20-year model. This will be true if the reversion value in Year 5 was correctly calculated.

In general, therefore, the length of the financial pro forma ought to be chosen based on the horizon over which you have useful information about the timing of specific cash flows. Beyond some date, your best forecast of cash flows may well be to assume a constant growth rate. As mentioned above, this is the same condition that makes using an exit cap rate appropriate. Thus, in practice, investment property pro formas are typically estimated over 5 or 10 years and are not necessarily tied to a commitment to a specific holding period. Depending on the investment, 5 or 10 years is a reasonable horizon to create detailed cash flow forecasts, and therefore beyond that time it may be appropriate to estimate the additional future value with an exit cap rate assumption.

Property valuation 31

Avoiding typical mistakes

When completing a cash flow pro forma on a prospective investment property, it is essential to create unbiased forecasts. An unbiased forecast means that any future deviation from your forecast should be a surprise. In the pro forma model for our industrial building, we have forecasted Year 3 total property cash flow to be $17,280. In 3 years, we will realize some particular level of total property cash flow. At that time, we should be surprised if Year 3 cash flow is higher than $17,280, and we should be equally surprised if cash flow is lower than $17,280. This is what it means to have an unbiased forecast.

Experience suggests that many property investors do not create unbiased pro formas of future cash flows. First, investors tend to overestimate cash flows, thinking that they will be able to increase rents and cut expenses more than it turns out that they are able to. In general, an investment property's cash flow growth should not be expected to exceed inflation for extended periods of time. This reflects the fact that property quality deteriorates as the property ages. Overestimation of future cash flow also may arise because investors underestimate the likelihood of economic downturns and the impact of such events on a particular property's tenants. Second, investors tend to systematically underestimate the level of capital expenditures needed to keep the property in a condition to lease effectively. Analysis of historical data on high quality property suitable for institutional investors indicates that between 20% and 40% of NOI is spent on capital expenditures on average. Although lower quality property typically involves lower levels of these expenditures, an investor would be wise to understand that these costs are considerable. If these were the only 2 mistakes that property investors tended to make, one would assume that would lead investors to overvalue property. Fortunately, or unfortunately depending on your perspective, the typical property investor makes a third mistake, by assuming an expected rate of return (discount rate) that also turns out to be higher than is realized. This tends to cancel the valuation impact of the first 2 mistakes.

Decision making

In evaluating the potential purchase of an investment property by conducting a discounted cash flow analysis, the investment decision becomes an easy one. If you can acquire the property for less than your estimated value, then the investment would be a good one. If you must pay more than your value estimate to acquire the property, then the investment would be best to avoid. This is the **net present value rule** applied to the decision to purchase an investment property.

Completing an unbiased discounted cash flow analysis of an investment property as described in this chapter is not easy. Many investors do not bother. Instead, they approach the investment decision somewhat differently. They estimate the future cash flows that the property will deliver – often making the mistakes outlined here – which leads to the cash flows being overly optimistic. Then, instead of thinking carefully about the rate of return the investment's risks imply, they instead ask, "What is the lowest price at which I can acquire this property?" By answering that question, they then calculate an internal rate of return (IRR) estimate for the investment. Finally, they ask themselves, "Am I happy with this return estimate?" They often answer this question without serious consideration of the risks associated with the investment. This IRR-based decision making is popular for property investment because the alternative – calculating an appropriate risk-adjusted rate of return for any particular investment – is particularly challenging. Nevertheless, only by considering the

Investment property valuation: golden opportunity

In the Golden Opportunity case, you are put in the shoes of Aurelia Dimas, the managing director of fixed income at Orrington Financial Partners. The case is set in 2010, shortly after the state of California offered for sale the Golden State Portfolio, a group of 11 state-owned office buildings. The portfolio was being offered as a sale-leaseback, meaning that along with the sale, the state was simultaneously signing a lease to rent the space from the new owners for 20 years. The case provides details on each of the 11 properties, information on each market, the terms of the state's proposed leases, and a financial pro forma prepared by the state's sales broker, CBRE. Dimas (you) must decide how much to bid for each property being offered. After reading the case, you will be able to:

1. Evaluate the cash flow projections for investment property that are prepared by others.
2. Understand how both theory and practical considerations influence the appropriate valuation of an investment property.

Golden opportunity: commercial real estate valuation

Aurelia Dimas, managing director of fixed income at Orrington Financial Partners, had spent the first week of April 2010 on a whirlwind tour of California. Her itinerary had not included Yosemite or Disneyland, however – she had been investigating the 11 properties in the Golden State Portfolio that were being offered as a sale-leaseback by the state of California.

Looking for a quiet place to think, Dimas drove to the top of Twin Peaks to enjoy the view of the Golden Gate Bridge to the north and the skyline of San Francisco's business district to the east. After taking a deep breath of fresh air, she took her tablet from her backpack and began to prepare her presentation for the next day's video conference with her CEO, Hank Christofferson.

Orrington Financial Partners had recently expanded its fixed-income portfolio to include real estate, and the offerings in the Golden State Portfolio seemed like a perfect fit – they provided diversification and stability over a period of decades. With the bid deadline of April 14 rapidly approaching, Dimas did not have much time to prepare her recommendations.

Background

Dimas's interest in real estate had come out of an unexpected challenge. While she was in college her family's home had been slated for demolition to make way for a new stadium, and Dimas had spent a significant amount of time researching the property and the future construction in order to accurately value the land. This experience inspired her to earn her MBA and pursue majors in real estate and finance.

She spent the next decade working for a property management company in Miami, where she successfully turned around poorly managed buildings throughout the southeastern United States. Her understanding of operations and finance led Christofferson to recruit Dimas in 2005 to head a new real estate group at Chicago-based Orrington Financial, an investment management firm with approximately 50 employees.

When Dimas first approached Christofferson with the opportunity to bid on the Golden State Portfolio, his eyes lit up. As described by the California Department of General Services

(Figure 2.1), the offering's size and scope were unprecedented. Christofferson tasked Dimas with visiting the properties and developing the company's bid. As Dimas examined the details of the portfolio's properties, she enjoyed thinking of herself as a modern-day prospector sifting for valuable nuggets of real estate (Figure 2.2).

California State Office Portfolio Hits the Market

Acquisition of Green Offices Viewed as a "Generational" Opportunity for Investment Capital

SACRAMENTO, Calif. – The California Department of General Services announced today that eleven state office properties, totaling nearly 7.3 million rentable square feet, are now on the market and expected to draw significant interest from capital investors worldwide, potentially eliciting offers in excess of $2 billion for the state. DGS's broker, CB Richard Ellis Group, Inc., has listed the properties online. Offers are due by April 14, 2010.

In June, Governor Schwarzenegger and the legislature authorized the sale of the properties located in Los Angeles, Oakland, Sacramento, San Francisco, and Santa Rosa. Once sold, the state anticipates retiring more than $1 billion in bond debt, saving California hundreds of millions in interest payments over the next two decades. The sale is also expected to net at least $660 million in proceeds that will be funneled directly into the General Fund, helping to save Californians from increased taxes and deeper cuts in state programs and services. "This sale will allow California to pay off debt, tap equity, and lock in some of the lowest rental rates seen in years," said DGS Acting Director Ron Diedrich. "The short- and long-term financial gains will be real to help shore up the state budget in the years to come."

The eleven state office properties are among some of California's most energy efficient and environmentally friendly, making the properties attractive to a market that is seeking sustainable, green designs. The U.S. Green Buildings Council's Leadership in Energy and Environmental Design (LEED) certification has been achieved on nearly all of the buildings.

"What the real estate investment community is looking for in today's market are secure, low-risk investment opportunities – occupied buildings with long-term, credit-worthy tenants, as well as increasingly 'green' product, both of which the state's portfolio offers," said Kevin Shannon, Vice Chairman for CB Richard Ellis and the lead broker on the sale-leaseback assignment. "We are confident that the expansive global marketing campaign we're launching today will attract strong national and international interest in this generational acquisition opportunity."

"California should not be in the volatile real estate business," said Diedrich. "As we lease these properties for the next twenty years, we can predictably budget our costs, knowing that the state will no longer be liable for unforeseen repair costs that are inherent in owning real estate."

Source: California Department of General Services press release, February 26, 2010.

Figure 2.1 Press release of portfolio sale

Region		Property Name	Address	Rentable Area (sq. ft.)	# of Floors	Year Built	LEED Certification	Projected Year 1 NOI
Bay Area	1	Public Utilities Commission Building	505 Van Ness Avenue San Francisco	270,768	5	1984	LEED Silver	$6,098,050
	2	San Francisco Civic Center	350 McAllister Avenue & 455 Golden Gate Avenue San Francisco	912,387	6 & 14	1922 & 1999	LEED Gold (1 bldg)	$22,040,256
	3	Elihu Harris Building	1515 Clay Street Oakland	700,589	24	1998	LEED Certified	$12,613,763
	4	Judge Rattigan Building	50 D Street Santa Rosa	92,368	4	1983	Registered	$1,040,445
Los Angeles	5	Junipero Serra State Building	320 West 4th Street Los Angeles	431,856	10	1914, 1999 (Renovated)	Registered (w/ certification goal of "Silver")	$6,799,418
	6	Ronald Reagan State Building	300 South Spring Street Los Angeles	739,158	14 & 16	1989	Registered (w/ certification goal of "Silver")	$12,195,530
Sacramento	7	Attorney General Building	1300 I Street Sacramento	376,866	17	1995	LEED Gold	$9,708,584
	8	Capitol Area East End Complex	1430 N Street; 1500, 1501,1615, and 1616 Capitol Avenue Sacramento	1,474,705	6 & 7	2002 & 2003	LEED Platinum (1 bldg), LEED Gold (4 bldgs)	$35,543,577
	9	Department of Justice Building	4949 Broadway Sacramento	381,718	2	1982	Registered	$4,936,426
	10	Franchise Tax Board Complex	9645 Butterfield Way Sacramento	1,814,056	1 to 4	1984, 1993, 2003 & 2005	LEED Gold (4 bldgs), LEED Silver (2 bldgs)	$34,310,182
	11	Cal EMA	3650 Schriever Avenue Rancho Cordova	116,687	1 & 2	2002	Registered	$2,921,246
		Total		7,311,158			Total	$148,207,477

Figure 2.2 Portfolio summary

Source: CBRE Golden State Portfolio Offering Memorandum.

State of California

While visiting the state capitol in Sacramento, Dimas met with Luke Orville, the longtime staff director of the legislative budget office, to learn more about the state's operations and the conditions that led to the sale-leaseback offering. Orville estimated that the state of California faced a budget deficit of nearly $20 billion for the 2009–2010 budget cycle. Governor Schwarzenegger had called for cuts in practically every aspect of government spending, which led to the proposed sale-leaseback of state-owned properties in the Golden State Portfolio. The transaction would raise funds, narrow the budget deficit, and reduce the risks of property ownership by reducing the state's significant holdings. Despite its budget difficulties, California had maintained a high bond rating that allowed the issuance of debt at favorable rates (Table 2.7).

The Portfolio

At the end of their conversation, Orville referred Dimas to Lea Takanaka, an agent for CB Richard Ellis (CBRE), which acted as the real estate broker for the Golden State Portfolio.

Table 2.7 Market data

	Yield
Treasury Bills	
3-month	0.13
6-month	0.19
1-year	0.32
2-year	0.80
5-year	2.28
10-year	3.61
30-year	4.56
Corporate Bonds	
Aaa	5.22
Baa	6.26
Municipal Bonds	
State and local bonds	4.34

Source: U.S. Department of the Treasury, Federal Reserve, March 1, 2010.

California Municipal Bonds Rating History

Year	Standard & Poor's	Fitch	Moody's
2000	AA	AA	Aa2
2001	A+	AA	Aa3/A1
2002	A	AA	A1
2003	BBB	BBB	A2/A3/Baa1
2004	A	A-	A3
2005	A	A	A2
2006	A+	A+	A1
2009	A	A/A-/BBB	A2/Baa1
2010	A-	A-	A1

Source: California Department of Finance.

36 *Property valuation*

CBRE had been selected based on the firm's experience with high-value properties and its fee of less than 0.5%.[1]

Takanaka explained to Dimas that CBRE was one of the world's largest global commercial real estate companies, formed when a spinoff from Coldwell Banker acquired the international holdings of London-based Richard Ellis International. As of 2010, the company had more than 400 offices in 60 countries, with 2.9 billion square feet under management.[2] CBRE maintained a wide range of real estate offerings, including "strategic advice and execution for property sales and leasing; corporate services; property, facilities, and project management; mortgage banking; appraisal and valuation; development services; investment management; and research and consulting."[3]

The Golden State Portfolio consisted of 11 properties totaling 7.3 million square feet in the markets of the San Francisco Bay Area, Los Angeles, and Sacramento. (See Figure 2.3 for a map and Figure 2.4 for an overview of each of the cities involved.) The buildings ranged in size from the 92,000-square-foot Rattigan Building in Santa Rosa to the 1.8-million-square-foot Franchise Tax Board Complex building in Sacramento. The majority of the buildings were LEED certified, which Takanaka argued would result in below-average energy costs.[4] In addition, the cost of construction had been paid by the state of California, so there were no current mortgages associated with the buildings.

In addition to sharing her personal experiences in the local market, Takanaka provided data on returns (Figure 2.5), sales (Figure 2.6), and operations (Table 2.8) over the past decade. Although Takanaka painted a sunny picture for future prospects, Dimas hesitated to rely on just 1 source of data; she also had a survey of local real estate investors from her favorite trade magazine (Table 2.9). In addition, Dimas asked an associate at Orrington Financial to compile a thorough list of recent office sales (Table 2.10).

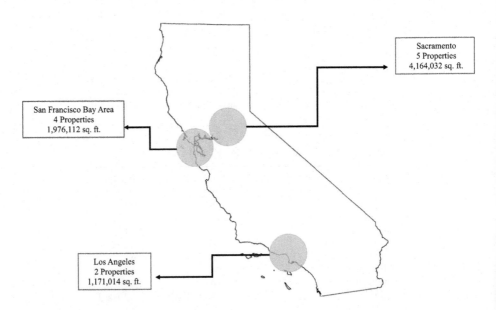

Figure 2.3 Portfolio map

SAN FRANCISCO

San Francisco has the unique distinction of being the most innovative city in America. The city is filled with a concentration of research organizations, an educated workforce, and a deep venture capital pool. San Francisco is also home to 29 Fortune 500 companies and 32 of *Inc.* magazine's fastest-growing private companies. There are also 6 leading research universities in close proximity to the city, including Stanford University and the University of California at Berkeley, and 5 national laboratories.

San Francisco is an excellent city to call home. The city was ranked number one in the Gallup-Healthways Well-Being Index, which measures aspects such as life evaluation, emotional and physical health, healthy behavior, work environment, and basic access. San Francisco is divided into a diverse set of districts, each with a unique character, including Fisherman's Wharf, Ghirardelli Square, Pier 39, and the famously crooked Lombard Street.

OAKLAND

Oakland is located in the East Bay region, which contains one of the most significant bioscience clusters in the nation. Influential firms such as Novartis, Bayer, Bio-Rad, Applied Biosystems, and Cell Genesys are based in this region. Oakland is also in close proximity to outstanding research institutions, including the Lawrence Livermore National Laboratory and Sandia National Laboratories. Both the commercial and government sectors are fueled by graduates of the University of California at Berkeley, located just north of the city. Additionally, Oakland is the only city in California to boast all three major sports franchises – the Oakland Raiders, Oakland A's, and Golden State Warriors. It is also the city where such culinary favorites as Rocky Road ice cream and the Mai Tai cocktail were created.

Figure 2.4 California submarket overview

Source: Picture credits: Cable car: Getty Image 498187957 LA Skyline: Getty Image 536191349

SANTA ROSA

Santa Rosa is located in the North Bay region, home to the world-renowned California wine country. Santa Rosa, the largest city in the region, has become the central business hub as a result of the booming wine and tourism industry. Other sectors with a significant presence include retail, technology, and medical, with major employers such as Kaiser Permanente, Agilent Technologies, and Medtronic.

Growth in the area is expected to accelerate with the planned completion in 2014 of the 70-mile Sonoma-Marin Area Rail Transit (SMART). A full range of educational institutions serves the city's needs, including Sonoma State University, Santa Rosa Junior College, and a top-rated public school system. Santa Rosa also offers diverse cultural attractions. The Wells Fargo Center for the Arts hosts performances ranging from music to comedy, many of which are broadcast as HBO specials. The Sonoma County Museum is the cornerstone of the Santa Rosa Arts District, which hosts a variety of events throughout the year.

SACRAMENTO

Sacramento is the state of California's capital, located in the beautiful Central Valley. It is regularly mentioned in lists of the country's most desirable places to live (number 8 by Kiplinger in 2008) and lies within a 90-minute drive of favorite destinations such as San Francisco, Napa Valley, and Lake Tahoe. Sacramento has a stable, recession-resistant economy due to government employment.

LOS ANGELES

Los Angeles, or the "City of Angels," is home to 72 miles of coastline, Hollywood's movie studios, and, of course, the rich and famous. LA is also the second most populous city in the United States, with 17.8 million people. Attractions along the coast include Malibu, Santa Monica, and Venice's Muscle Beach. For a greener setting, there is the massive Griffith Park, which features the LA Observatory and the LA Zoo. No visit to Los Angeles would be complete without a stop in Hollywood. While strolling through Hollywood, visitors can find the Walk of Fame, well-known theaters, and the site of the annual Oscar Awards. However, LA offers more than just glamour, as the city is headquarters to 16 Fortune 500 companies, which anchor a diverse economy that includes aerospace, textiles, trade, biotech, and entertainment.

Figure 2.4 (Continued)

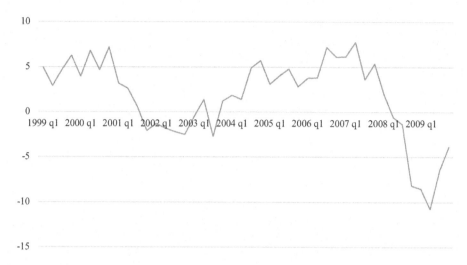

Figure 2.5 Returns to office property in San Francisco (percentage points)
Source: National Council of Real Estate Investment Fiduciaries.

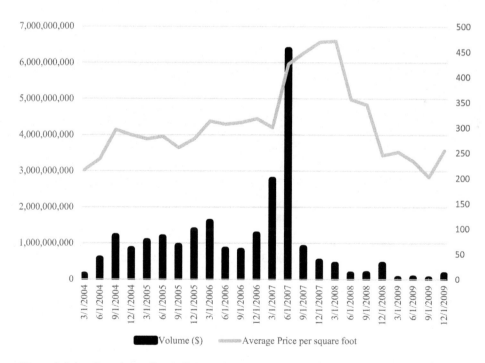

Figure 2.6 San Francisco office building sales
Source: Real Capital Analytics, Inc.

Table 2.8 San Francisco – operating statistics for Class A office market

Period	Number of Buildings	Total Rentable Building Area (sq. ft.)	Vacancy (%)	Net Absorption (sq. ft.)	Quotes Rates ($ per sq. ft. per year)
2009 4Q	313	73,360,544	13.2	(131,548)	30.35
2009 3Q	313	73,360,544	13.0	81,947	31.65
2009 2Q	312	73,202,615	13.0	(997,906)	31.55
2009 1Q	312	73,202,615	11.6	(452,574)	34.54
2008 4Q	312	73,202,615	11.0	(70,419)	37.87
2008 3Q	308	72,299,956	9.8	(109,933)	41.05
2008 2Q	307	71,743,637	8.9	404,320	42.21
2008 1Q	301	70,614,715	8.0	426,320	40.74
2007	301	70,614,715	8.6	2,146,959	40.86
2006	300	70,458,715	11.5	1,878,276	33.07
2005	299	70,326,589	14.0	3,305,948	28.27
2004	295	69,498,657	17.7	1,617,259	26.47
2003	295	69,498,657	20.0	858,420	25.78
2002	283	67,362,927	18.8	154,957	28.28
2001	271	64,791,590	15.8	3,431,551	36.14
2000	248	60,389,989	4.0	1,801,632	68.24

Source: CoStar Group, Year-End 2009.

Table 2.9 San Francisco – real estate investor survey for the office market

	Current Quarter	Last Quarter	Previous Year
Discount rate (IRR)[a]			
Range (%)	8.00–14.00	8.00–14.00	7.00–12.00
Average (%)	9.74	9.75	8.16
Change (basis points)	–	(1)	158
Overall cap rate (OAR)[a, b]			
Range (%)	6.00–11.00	6.00–11.00	4.50–9.00
Average (%)	7.94	7.84	6.16
Change (basis points)	–	10	178
Residual cap rate[c]			
Range (%)	6.00–12.00	6.25–12.00	5.00–9.00
Average (%)	8.14	8.13	7.07
Change (basis points)	–	1	107
Market rent change rate[d]			
Range (%)	(20.00) – 3.00	(20.00) – 3.00	0.00–10.00
Average (%)	(1.00)	(2.06)	2.73
Change (basis points)	–	106	(373)
Expense change rate[d]			
Range (%)	0.00–3.00	0.00–3.00	2.00–5.00
Average (%)	2.67	2.67	3.10
Change (basis points)	–	0.00	(43)
Average marketing time[e]			
Range	1.00–12.00	1.00–12.00	1.00–12.00
Average	6.89	7.71	6.33
Change (%)	–	(10.64)	8.85

Source: Korpacz Real Estate Investor Survey, Fourth Quarter 2009.

a Rate on unleveraged, all-cash transactions.
b Also known as the direct cap rate.
c Also known as the exit cap rate or terminal cap rate.
d Initial rate of change.
e In months.

Table 2.10 Recent office sales in San Francisco

Address	Date	Square Feet	Year Built	Year Renovated	Price ($)	Price per Sq. Ft. ($)	Cap Rate (%)	Occupancy at Sale (%)
538–552 Montgomery St	Feb-10	93,006	1931	—	12,650,000	136	—	—
1 Ecker Pl	Jan-10	60,000	1906	1985	14,000,000	233	—	—
49 Stevenson St	Jan-10	121,186	1989	—	24,200,000	200	4.7	80
211 Main St	Dec-09	371,000	1973	1998	111,766,667	301	8.5	100
921 Howard St	Dec-09	28,534	1900	—	9,700,000	340	—	—
640 2nd St	Dec-09	22,712	1925	2000	5,900,000	260	—	0
550 Terry Francois St	Nov-09	283,000	2002	—	135,500,000	479	10.3	100
562–566 Market St	Sep-09	64,955	1923	—	9,200,000	142	—	75
239 Grant Ave	Jun-09	11,000	1932	—	8,250,000	750	7.5	100
250 Montgomery St	Jun-09	116,078	1986	—	21,240,000	183	—	50
731 Sansome St	Jun-09	37,811	1911	1989	7,000,000	185	—	—
717 Battery St	Feb-09	32,400	1907	1968	13,000,000	401	—	0
100 Harrison St	Feb-09	150,000	1942	—	14,000,000	93	—	—
706 Sansome St	Nov-08	12,000	1909	2008	6,800,000	567	—	—
815 Hyde St	Oct-08	21,607	1904	2001	5,800,000	268	6.9	100
235 Pine St	Oct-08	149,147	1991	—	56,500,000	379	—	—
760 Market St	Aug-08	267,446	1908	—	130,000,000	486	—	78
470–492 Pacific Ave	Aug-08	16,644	1907	—	6,400,000	385	—	—
410 Pacific Ave	Jun-08	24,661	—	—	13,500,000	547	—	100
55 Francisco St	Jun-08	143,695	1916	1983	58,000,000	404	5.8	95
500 Terry Francois Blvd	Jun-08	291,000	2007	—	149,000,000	512	—	0
1 Montgomery St	Apr-08	75,880	1908	1985	36,000,000	474	—	100
1095–1097 Market St	Apr-08	59,794	1905	—	9,000,000	151	—	—
1 Montgomery St	Mar-08	75,880	1908	1985	8,799,023	116	—	100
1 Beach St	Mar-08	97,000	1971	—	27,000,000	278	—	99
1 Beach St	Mar-08	97,000	1971	—	18,000,000	186	—	93
166 Geary St	Feb-08	60,000	1907	—	22,500,000	375	6.0	100
944 Market St	Feb-08	45,465	1907	—	6,000,000	132	5.0	100
816–818 Mission St	Feb-08	28,516	1907	—	15,000,000	526	—	—
100–150 Van Ness Ave	Jan-08	597,574	1974	—	118,500,000	198	8.4	75
490 Post St	Jan-08	45,000	—	—	6,700,000	149	—	91
250 4th St	Jan-08	29,590	1947	—	9,000,001	304	—	100
199 Fremont St	Jan-08	401,043	2000	—	275,000,000	686	5.2	97

Source: Real Capital Analytics, Inc.

42 Property valuation

Preparing the presentation

Dimas began to compile her analysis for presentation to Christofferson. She knew it would be important to start by explaining exactly what was involved in a sale-leaseback. There were many reasons organizations might want to enter into a sale-leaseback agreement: to convert equity into cash, to have an alternative to conventional financing, to gain favorable financing terms, to improve the balance sheet with a current rather than fixed asset, and to avoid debt restrictions.[5] Strategically, organizations sometimes wanted to divest their real estate holdings in order to focus on activities that were more central to their missions.

Of course, a sale-leaseback arrangement was viable only if it provided a stable, long-term investment to the buyer, typically in the form of high rental rates and a lengthy contract as compensation for assuming the risk of a default on the lease. A seller, on the other hand, assumed the risk of a buyer default and change of ownership that could result in new lease terms or even eviction, in addition to the loss of flexibility in subleasing and building modification.

With so many properties in the Golden State Portfolio to value, Dimas needed a summary (Figure 2.7) of each one to keep track of all the details. She also had a lease abstract (Figure 2.8) for each building, which included details she needed to estimate cash flows to be paid and received by the new owner. Takanaka had also provided a pro forma developed by CBRE for each property (Figure 2.9).

Based on her valuation, Dimas needed to recommend how to bid on the portfolio. One option was to bid separately for each property she considered a good investment. The other possibility was to team up with another firm to bid on the entire portfolio.

Property:

   Civic Center
   350 McAllister Street and 455 Golden Gate Avenue
   San Francisco, CA 94102

LEED Certification:

   LEED Gold (455 Golden Gate Avenue)

Year Built:

   350 McAllister Street – 1922 original construction with subsequent addition 1930; renovated 1999
   455 Golden Gate Avenue – 1998

Number of Structures:

   Two office buildings connected by an interior open atrium:
   350 McAllister Street – 6-story office building with garage
   455 Golden Gate Avenue – 14-story office building with subterranean garage and auditorium

---

Figure 2.7 Property summary of the San Francisco Civic Center

> ***Parcel Numbers:***
>
> Block 0765, Lots 002 and 003
>
> ***Site Area:***
>
> 350 McAllister Street – 252,018 square feet
> 455 Golden Gate Avenue – 807,879 square feet
>
> ***Total Rental Area:***
>
> 912,387 square feet
>
> ***Zoning:***
>
> 350 McAllister Street – Public
> 455 Golden Gate Avenue – Public
>
> ***Number of Floors:***
>
> 350 McAllister Street – 6
> 455 Golden Gate Avenue – 14
>
> ***Site Location:***
>
> 350 McAllister Street and 455 Golden Gate Avenue are both located between Polk and Larkin streets in downtown San Francisco. Adjacent properties consist of similar mid-rise and high-rise civic center buildings, including San Francisco City Hall to the southwest.
>
> *Source*: CBRE Golden State Portfolio Offering Memorandum.

Figure 2.7 (Continued)

Dimas needed to get this right – an investment in the Golden State Portfolio would represent a significant increase in the holdings of Orrington Financial. As she opened her spreadsheet program to begin her valuation analysis, she remembered a cautionary remark her real estate finance professor had often repeated: "Skilled financial analysts can make a spreadsheet justify anything – so think carefully about your assumptions." She had already started thinking about her own assumptions for the property financials and needed to finalize them. However, even with all of her experience, she had never dealt with a government seller and the nonmarket factors associated with its motivation to enter into a sale-leaseback.

As the sun was setting, Dimas put her tablet into her backpack and prepared to drive to her hotel. She had a video conference the next day with her CEO and a presentation to finish.

Lease term	20 years, July 2010 to June 2030
Rentable space	912,387 square feet (1500 square feet reserved for property management office)
Rent	Years SF $/Month $/SF/Month $/Year $/SF/Year
	1 to 5 910,887 $2,968,630 $3.26 $35,623,560 $39.11
	6 to 10 910,887 $3,265,493 $3.58 $39,185,916 $43.02
	11 to 15 910,887 $3,592,042 $3.94 $43,104,504 $47.32
	16 to 20 910,887 $3,951,246 $4.34 $47,414,952 $52.05
Growth assumptions	Market rent – 3%; CPI – 3%; Operating expenses – 3%; Property taxes – 2%
Parking	Exclusive right (but no obligation) to the 65 parking spaces at a charge of $250 per stall per month. Rate increases by 10% every 5 years.
Operating expenses	Provided at Lessor's sole cost and expense. Includes security, cleaning, repairs and maintenance, utilities (inclusive of sewer, trash, and water but exclusive of gas and electricity), management, lot and landscaping, parking, taxes, and insurance.
CPI escalator	Every 12 months, the monthly rent will automatically increase by 1/12th of an amount determined by multiplying $9,971,440[1] by the percentage increase in the CPI index for the preceding 12 months.
Real estate tax escalator	Every 12 months, monthly rent will automatically increase by 1/12th of an amount determined by multiplying the annual property tax expense for the preceding 12 months by the percent increased capped at 2%. Current real estate tax millage rate is 1.14%. Note: Buyer is responsible for adjusting property taxes bases upon offer price.
Painting and carpet	Provided at Lessor's sole cost and expense. Repainting to occur every 5 years at $2 per square foot. Carpets replaced every 10 years at $2 per square foot.
Subletting	With written consent of Lessor, State may sublet the premises.
Insurance	Provided at Lessor's sole cost and expense as required by law.
Right of first refusal	Eligible for offers tenured 90 days or more prior to termination date. The State shall have 30 days from receipt of the offer to acquire the Property under the same terms and conditions.
Option to renew lease	State shall have the option to extend the lease for 6 additional terms of 5 years. State shall provide written notice at least 24 months prior to expiration.
Onsite management	Provided at the Lessor's sole cost and expense at $279,996 based on a portfolio sale.

Figure 2.8 Lease abstract for San Francisco Civic Center

Source: CBRE Golden State Portfolio Offering Memorandum.

Note: [1] $9,971,440 represents the sum total of all operating expenses other than Real Estate Taxes.

	FY 2011 $/SF/YR [2]										
Physical occupancy		100.00%	100.00%	100.00%	100.00%	100.00%	100.00%	100.00%	100.00%	100.00%	100.00%
Overall economic occupancy [1]		100.00%	100.00%	100.00%	100.00%	100.00%	100.00%	100.00%	100.00%	100.00%	100.00%
Total operating expenses PSF per year		$15.17	$15.58	$16.00	$16.44	$16.89	$17.35	$17.82	$18.31	$18.81	$19.32	$19.85
Revenues												
Scheduled base rent												
Gross potential rent	$39.11	$35,682,218	$35,682,218	$35,682,218	$35,682,218	$35,682,218	$39,250,440	$39,250,440	$39,250,440	$39,250,440	$43,175,484
Absorption & turnover Vacancy	0	0	0	0	0	0	0	0	0	0	0
Base rent abatements	0	0	0	0	0	0	0	0	0	0	0
Total scheduled base rent	39.11	35,682,218	35,682,218	35,682,218	35,682,218	35,682,218	39,250,440	39,250,440	39,250,440	39,250,440	43,175,484
Expense reimbursement	0	0	0	0	0	0	0	0	0	0	0
Parking revenue	0.21	195,000	195,000	195,000	195,000	195,000	214,500	214,500	214,500	214,500	235,950
State of CA expense increases	0	0	375,836	762,175	1,159,316	1,567,569	1,987,251	2,418,687	2,862,215	3,318,179	3,786,935	4,268,849
Total gross revenue	39.32	35,877,218	36,253,054	36,639,393	37,036,534	37,444,787	41,452,191	41,883,627	42,327,155	42,783,119	43,251,875	47,680,283
General vacancy loss	0	0	0	0	0	0	0	0	0	0	0
Effective gross revenue	39.32	35,877,218	36,253,054	36,639,393	37,036,534	37,444,787	41,452,191	41,883,627	42,327,155	42,783,119	43,251,875	47,680,283
Operating expenses												
Security	(1.30)	(1,184,400)	(1,219,932)	(1,256,530)	(1,294,226)	(1,333,053)	(1,373,044)	(1,414,236)	(1,456,663)	(1,500,362)	(1,545,373)	(1,591,735)
Cleaning	(3.80)	(3,467,070)	(3,571,082)	(3,678,215)	(3,788,561)	(3,902,218)	(4,019,284)	(4,139,863)	(4,264,059)	(4,391,981)	(4,523,740)	(4,659,452)
Repairs & maintenance	(2.53)	(2,304,311)	(2,373,440)	(2,444,644)	(2,517,983)	(2,593,522)	(2,671,328)	(2,751,468)	(2,834,012)	(2,919,032)	(3,006,603)	(3,096,801)
Utilities	(0.20)	(182,400)	(187,872)	(193,508)	(199,313)	(205,293)	(211,452)	(217,795)	(224,329)	(231,059)	(237,991)	(245,130)
Management fee	(0.31)	(279,996)	(288,396)	(297,048)	(305,959)	(315,138)	(324,592)	(334,330)	(344,360)	(354,691)	(365,331)	(376,291)
Onsite office expense	(0.22)	(205,063)	(211,215)	(217,551)	(224,078)	(230,800)	(237,724)	(244,856)	(252,202)	(259,768)	(267,561)	(275,588)
Admin (Excl mgt fee)	(0.47)	(426,000)	(438,780)	(451,943)	(465,502)	(479,467)	(493,851)	(508,666)	(523,926)	(539,644)	(555,833)	(572,508)
Lot & landscaping	(0.28)	(256,800)	(264,504)	(272,439)	(280,612)	(289,031)	(297,702)	(306,633)	(315,832)	(325,307)	(335,066)	(345,118)
Parking	(0.13)	(121,200)	(124,836)	(128,581)	(132,439)	(136,412)	(140,504)	(144,719)	(149,061)	(153,533)	(158,139)	(162,883)
Real estate taxes	(4.24)	(3,865,522)	(3,942,832)	(4,021,689)	(4,102,123)	(4,184,164)	(4,267,849)	(4,353,206)	(4,440,270)	(4,529,075)	(4,619,657)	(4,712,050)
Insurance	(1.69)	(1,544,200)	(1,590,526)	(1,638,242)	(1,687,389)	(1,738,011)	(1,790,151)	(1,843,856)	(1,899,171)	(1,956,146)	(2,014,831)	(2,075,276)

Figure 2.9 San Francisco Civic Center Cash Flow Projections

Source: CBRE Golden State Portfolio Offering Memorandum.

Physical occupancy		100.00%	100.00%	100.00%	100.00%	100.00%	100.00%	100.00%	100.00%
Total operating expenses	(15.17)	(13,836,962)	(14,213,415)	(14,600,390)	(14,998,185)	(15,407,110)	(15,827,481)	(16,259,628)	(16,703,885)	(17,160,598)	(17,630,125)	(18,112,832)
Net operating income	24.16	22,040,256	22,039,639	22,039,003	22,038,349	22,037,677	25,624,710	25,623,999	25,623,270	25,622,521	25,621,750	29,567,451
Capital costs										
Tenant improvements	0.00	0	0	0	0	0	0	0	0	0	0	0
Leasing commissions	0.00	0	0	0	0	0	0	0	0	0	0	0
Capital reserves	(0.15)	(136,858)	(140,964)	(145,193)	(149,548)	(154,035)	(158,656)	(163,416)	(168,318)	(173,368)	(178,569)	(183,926)
State of CA – Paint & carpet [3]	0.00	0	0	0	0	0	(1,821,774)	0	0	0	0	(3,643,548)
Total capital costs	(0.15)	(136,858)	(140,964)	(145,193)	(149,548)	(154,035)	(1,980,430)	(163,416)	(168,318)	(173,368)	(178,569)	(3,827,474)
Operating cash flow	$24.01	$21,903,398	$21,898,675	$21,893,810	$21,888,801	$21,883,642	$23,644,280	$25,460,583	$25,454,952	$25,449,153	$25,443,181	$25,739,977

[1] This figure takes into account vacancy/credit loss, absorption vacancy, turnover vacancy, and base rent abatements.
[2] Based on 912,387 square feet.
[3] State of California to have premises repainted every 5 years (estimated to be $2 PSF) and re-carpeted every 10 years (estimated to be $2 PSF).

Figure 2.9 (Continued)

Appendix: the relationship between cap rates and property returns

The return on a risky asset can be deconstructed into a cash flow yield and a capital gain. We can use this intuition to express the returns to investing in property. Defining r to be the return on an investment property, $CAPEX$ to be the capital expenditures of the property, and P to be the property's price, we can express the financial return to a property investment at time t as:

$$r_t = \frac{NOI_t - CAPEX_t + P_t}{P_{t-1}} - 1 = \text{cash flow yield} + \text{capital gain rate}$$

We can then substitute for the price, P, based on our definition of a cap rate c. This yields:

$$r_t = \frac{NOI_t - CAPEX_t}{\frac{NOI_t}{c_{t-1}}} + \frac{\frac{NOI_{t+1}}{NOI_t}}{\frac{NOI_t}{c_{t-1}}} - 1$$

$$r_t = c_{t-1}\left(1 - \frac{CAPEX_t}{NOI_t}\right) + \frac{\frac{NOI_{t+1}}{NOI_t}}{\frac{c_t}{c_{t-1}}} - 1$$

Letting g represent growth rates and ϕ represent the fraction of NOI that is spent on capital expenditures, we can rewrite the expression for property returns as:

$$r_t = c_{t-1}(1 - \phi_t) + \frac{1 + g_{t+1}^{NOI}}{1 + g_t^c} - 1$$

This expression has several implications for the impact of the sources of investment on returns in real estate at time t. First, returns are higher whenever the previous period cap rate is higher. This makes sense because the previous period's cap rate is a measure of the income the property produces relative to the previous price of the property. Second, returns are higher when investment in capital expenditures is lower. This is literally true in the sense that lower expenditures translate into higher cash flows, although it neglects the long-run impact of such decisions on future growth rates. Third, returns are higher when NOI growth is higher. This is the foundation of much of the real estate investment industry – how can we increase NOI? Finally, returns are higher when the growth rate of the cap rate is negative. This means that an investor earns higher returns when cap rates decline. This also makes sense in that declining cap rates mean that investors in the market will pay more for each dollar of income the property is earning.

Having derived an expression that identifies the short-term drivers of real estate returns, it is perhaps most useful to consider an investor looking to the distant future and wondering what cap rates imply about returns. In this distant future, we would expect returns and cap rates to be constant. Further, we would assume NOI to grow at a constant rate and the property investor to make capital expenditures at a constant rate that is consistent with maximizing the value of the property. These long-run assumptions imply that our expression for investor returns can be rewritten as:

$$r = c(1 - \phi) + g^{NOI}$$

This equation gives us a relationship between cap rates and expected returns (discount rates) that must hold in the distant future. You may ask why this long-run relationship is relevant for property valuation. It is significant because it is a relationship that should hold within the assumptions of a model of property cash flow. As we have seen, reversion cash flow is typically modeled as being derived from an exit cap rate assumption. What this expression tells us is that the exit cap rate assumption should be tied to long-run assumptions regarding discount rates, NOI growth rates, and the fraction of NOI that must be spent on capital expenditures.

This relationship also gives us another validation check. Sometimes investors are presented with a cash flow pro forma prepared by others. In such a situation, one should check the assumptions about exit cap rates, discount rates, and growth rates. Often, investors treat all 3 factors as free parameters in a model, despite the fact that in the long run (at exit) they are related.

Notes

1 California Department of General Services press release, "California Signs Deal with Broker to Sell $2 Billion in Real Estate," December 11, 2009.
2 CB Richard Ellis Group, Inc., "Worldwide Business Activity 2010," www.cbre.com/EN/AboutUs/MediaCentre/Documents/ CBRE_Worldwide_2010.pdf.
3 CB Richard Ellis Worldwide, "Investor Relations," http://ir.cbre.com/phoenix.zhtml?c=176560&p=irol-irhome (accessed September 8, 2011).
4 CBRE Golden State Portfolio Offering Memorandum.
5 Donald J. Valachi, CCIM Institute, "Sale-Leaseback Solutions: Examine the Business and Tax Considerations of These Transactions," www.ccim.com/cire-magazine/articles/sale-leaseback-solutions (accessed September 8, 2011).

3 Debt financing

Introduction

Many real estate investors finance the purchase of property, in part, with borrowed money. In Chapter 3, we begin by describing the typical components of a **commercial mortgage**, which is the most common form of borrowing for investment property. We then turn to the **mortgage underwriting process** – the method by which a lender will evaluate what it is willing to lend to support your property acquisition. We conclude by arguing that the willingness to lend need not be the determining factor in deciding how much to borrow. By the end of this chapter, you will understand how to calculate how much cash flow you will receive from the property after making the required payments to the lender as well as how a lender will decide how much debt your property will support.

Key features of a borrowing contract

When borrowing to help finance an investment property, an investor will be required to agree to the terms dictated by the lender. In this section, we review the standard terms borrowers face.

Pledging the property as collateral

When lenders provide money to investors to help purchase investment property, they want to be repaid. Before providing funds, the lender will require the investor to sign 2 legal agreements – the promissory note and the mortgage. The **promissory note** states that the borrower is obligated to repay and specifies the terms that are expected of repayment. The **mortgage** pledges the underlying investment property as collateral to secure the investor's obligation to repay. Simply stated, the mortgage contract gives the lender the right to take possession of the underlying property – through a legal process known as foreclosure – if the investor fails to meet the obligations specified in the note. Knowing that they may lose possession of the property provides incentives for the investor to repay the loan as promised. Likewise, knowing that they can access the value of the underlying property as a source of repayment provides lenders with confidence to lend. For ease of exposition, when this book refers to a mortgage contract, it is a reference to *both* the actual mortgage and the accompanying promissory note.

Determinants of the required payments

The mortgage documents provide the details needed to calculate the payments that the borrower has promised to make to the lender. In particular, the note specifies (1) the amount

50 Debt financing

borrowed; (2) the interest rate, typically quoted as an annual percentage rate (APR); (3) the compounding interval, which is typically monthly and which we will assume to be equal to the payment frequency; (4) the maturity of the loan; and (5) the amortization period – that is, the time over which the loan balance would reach zero if only monthly payments were made – of the loan.

Of particular relevance for investors is the fact that the maturity of a mortgage on an investment property is generally *not* equal to its amortization period. This may strike you as somewhat unusual. Most residential mortgages, car loans, and other consumer loans are typically fully amortizing. This means that after making the number of monthly payments specified in the loan, the loan balance will be reduced to zero. For example, if you finance the purchase of a new car with a 5-year loan requiring monthly payments, the amount you owe to your lender will be reduced to 0 precisely after making the 60 (5 years × 12 months per year) required payments. Similarly, if you have a traditional 30-year fixed rate mortgage on your home, you will own 100% of your home after making 360 monthly payments. Although possible, it is generally *not* the case that mortgage loans secured by investment property are fully amortizing. That is, the loan will *mature*, meaning that the lender ends its current loan relationship with the borrower, when there still remains a substantial payment required by the borrower to the lender. This is because the amortization period of a commercial mortgage is generally longer than the maturity. Lenders want to have confidence in the property's ability to generate the cash flow sufficient to repay the loan. Therefore, if a property's tenants have leases that are shorter in duration than 5 years, then a lender may wish to have the loan mature in 5 years. Of course, in this circumstance, the loan could also be designed with an amortization period that is also 5 years. In practice, short amortization periods of investment property mortgages would imply unaffordable monthly payments. Thus, most investment property mortgages have an amortization period significantly longer than their maturity.

It is helpful to illustrate this with an example.

> Example 3.1: Suppose that our investor purchasing the 2-unit industrial building will finance this acquisition, in part, with a $3,000,000 mortgage loan from a local bank. According to the terms of the loan, the investor is required to make monthly payments based on an interest rate of 4.25% (APR) for 5 years, at which point the loan matures. However, the mortgage payments were established with an amortization period of 25 years. What are the monthly and annual payments? How much is still owed when the loan matures?

Monthly payments required on a loan can be calculated using Excel. In this case, the monthly payment required on this loan is $16,252.14, which is calculated in Excel as =PMT(4.25% ÷ 12, 25 × 12, 3000000). Note that the number of payments used in the formula is 300, reflecting monthly payments for 25 years (the amortization period). This means that if the borrower were allowed to make monthly payments of $16,252.14 for 300 consecutive months, the loan balance that began at $3,000,000 would be reduced to 0.[1] However, this loan has a maturity of 5 years. That means that after the investor has made 60 monthly payments, the bank requires the entire remaining balance to be repaid immediately (also at the end of the 60th month).

So, how much is still owed after 60 monthly payments? To calculate this amount, we use the **standard rules of amortization**. These rules are shown in Table 3.1.

Let's work through the first 2 payments to see how the loan balance amortizes. Beginning with Rule 1, we know that the initial mortgage balance is simply the amount borrowed,

Debt financing 51

Table 3.1 The rules of amortization

1 The initial outstanding principal balance equals the initial contract principal specified in the loan agreement.
2 The interest owed each period equals the applicable interest rate times the outstanding principal balance at the end of the previous period.
3 The principal amortized (paid down) in each payment equals the total payment less the interest owed.
4 The outstanding balance after each payment equals the previous outstanding balance minus the principal paid in the payment.

Table 3.2 Sample amortization table

Payment	Outstanding Balance	Monthly Debt Service	Interest	Monthly Principal
	$3,000,000.00			
1	$2,994,372.86	$16,252.14	$10,625.00	$5,627.14
2	$2,988,725.78	$16,252.14	$10,605.07	$5,647.07
3	$2,983,058.71	$16,252.14	$10,585.07	$5,667.07
	...			
59	$2,631,486.30	$16,252.14	$9,344.31	$6,907.83
60	$2,624,554.00	$16,252.14	$9,319.85	$6,932.30

which in this example is $3,000,000. Using Rule 2, the interest owed in the first month is equal to the monthly interest rate (4.25% ÷ 12) multiplied by the outstanding balance of $3,000,000. This equals $10,625.00. According to Rule 3, that means that $5627.14 of our loan (total payment of $16,252.14 less $10,625.00 of interest) is amortized in the first month. According to Rule 4, this means that the outstanding balance after the first monthly payment is reduced to $3,000,000 – $5627.14 = $2,994,372.86.

Let's repeat these calculations for the second month. Rule 2 allows us to calculate the interest owed as 4.25% ÷ 12 × $2,994,372.86 = $10,605.07. Rule 3 tells us that the principal amortized in the second month is equal to $5647.07 (= $16,252.14 – $10,605.07), which according to Rule 4 reduces the principal owed to $2,988,725.78 = $2,994,372.86 – $5647.07.[2] In a spreadsheet, this is easy to repeat for any number of payments. Table 3.2 details the information necessary to answer the question, "How much is still owed after 60 monthly payments?"

As calculated in Table 3.2, our property investor will still owe the lender $2,624,554.00 after making the 60 required monthly payments. Because the loan has matured, the investor will not only have to make the monthly payment in month 60 but also have to repay the remaining $2,624,554.00. In our example, we have assumed that the investor is selling the property at the same time, so the remaining loan balance can be paid from the sales proceeds. More generally, the lump-sum loan repayment might alternatively be paid from the proceeds of a new mortgage loan, which refinances the existing, maturing loan.

Once an entire amortization table is completed, it becomes straightforward to aggregate the monthly payments to the same, annual frequency in which we have been calculating cash flows. For instance, regular debt service in the first year is simply equal to 12 × $16,252.14 = $195,025.72. Debt service in Year 5, the year of reversion, is equal to the sum of the regular debt service ($195,025.72) and the outstanding balance remaining on the loan after 5 years ($2,624,554.00). The investor pays this debt service out of the property's cash flow (when possible), implying that the cash flow to equity (to the investor) can be calculated

52 Debt financing

as the difference between total property cash flow and **total debt service**. This is illustrated in Table 3.3 with an assumed purchase price of $4,600,000.

Order of application of payments

Mortgages specify how a lender will treat a payment from a borrower. Under normal circumstances – that is, when the borrower is making a scheduled payment on time – this is irrelevant. However, if a borrower is ever late or otherwise fails to make a promised payment on time, the loan documents may call for the borrower to be subject to various fees and the lender may incur additional expenses associated with getting the borrower to pay. In these circumstances, it is important that the mortgage specify how any payment ultimately received from a borrower is to be treated. Typically, cash flow from a borrower is first used to pay any extraordinary expenses that the lender incurred because the borrower defaulted on the loan. Next, the cash flow is used to pay any loan penalty fees. Third, the cash flow is used to pay interest due. Finally, any remaining cash flow reduces the outstanding principal balance of the loan.

Assumption

Some investment property mortgages are **assumable**. This means that an investor can assume (take over) the responsibility for an existing mortgage loan when they acquire a property from a seller. Typically, a property investor prefers the property seller to prepay their existing mortgage so that the investor can arrange its own financing. However, there are times when an investor would benefit from the ability to assume existing debt. For example, if credit conditions are tight, it may be difficult for an investor to borrow as much as is currently being

Table 3.3 Calculating cash flow to equity

	Year 0	Year 1	Year 2	Year 3	Year 4	Year 5
Net operating income (NOI)		$184,680	$405,240	$337,280	$481,460	$477,822
Building improvements		$0	$0	$80,000	$0	$0
TIs		$60,000	$0	$60,000	$0	$0
Leasing commissions		$165,750	$0	$180,000	$0	$0
Holding period cash flow		–$41,070	$405,240	$17,280	$481,460	$477,822
Reversion cash flow						
Sales price						$5,926,615
Sales commission						$177,798
Total property cash flow	–$4,600,000	–$41,070	$405,240	$17,280	$481,460	$6,226,639
Total debt service	–$3,000,000	$195,026	$195,026	$195,026	$195,026	$2,819,580
Total cash flow to equity	–$1,600,000	–$236,096	$210,214	–$177,746	$286,434	$3,407,059
Property return	9.52%					
Debt return	4.20%*					
Equity return	16.26%					

Note: Table 3.3 has added an additional column representing the acquisition of the property and how it was financed in Year 0. Doing so allows for the calculation of returns at the property, debt, and equity levels.

* You may recall that the mortgage had an interest rate of 4.25%. The reason that the debt returns are shown as 4.20% is because the cash flows have been annualized before returns were calculated. Calculating the return to debt at a monthly frequency would deliver the expected 4.25% return.

used. Assumption can also benefit an investor if it helps them avoid the restrictions or fees associated with prepayment discussed later. Assumption also potentially allows a seller to pass along a favorable interest rate to a new investor. Even when assumptions are allowed, the mortgage lender will have the ability to evaluate any risks presented by the new investor and to adjust terms as they deem necessary.

Recourse

Suppose an investor defaults on a mortgage and the lender forecloses on the property, yet when the lender sells the foreclosed property it still does not recover what it is owed. In this case, the amount still owed is called a **deficiency**. A loan with **recourse** is one where, in this circumstance, the investor would still be legally liable for the deficiency. A loan that is non-recourse is one where only the property serves as collateral for the mortgage. Non-recourse lenders cannot pursue investors for any deficiencies.[3]

All else equal, borrowers prefer loans that are non-recourse. However, all else is not typically equal. That is, investors may not be able to acquire financing without accepting recourse. In other instances, recourse loans may come with a lower interest rate than similarly structured non-recourse lending.

Likewise, you might have thought that lenders would always prefer recourse lending. This, too, turns out not to be the case. Recourse only arises in circumstances where the borrower has defaulted. In such a situation, lenders often prefer to quickly foreclose on the property, realize any loss, and move on. When a borrower faces recourse, however, they may tend to delay foreclosure by filing lawsuits against the lender, declaring personal bankruptcy, or pursuing a variety of other actions that would delay the lender's ultimate possession of the investment property. As a result, some lenders will only lend on a non-recourse basis because they believe that non-recourse borrowers will generally allow a lender to foreclose once it seems likely that there is little chance for the borrower to recover any value from the investment property.

Lockbox

As discussed in Chapter 2, a property investor will collect rent from the tenants in the property. In many cases, rents are simply paid to the property owner, who is then independently responsible for making payments on the mortgage loan. Some mortgage contracts require the use of a lockbox. A **lockbox** is a bank account established by the investor from which loan payments will be made. A hard lockbox agreement would require that the property's tenants pay rent into the lockbox account. Only after the required payment to the lender has been made would the investor be able to withdraw any excess funds. A soft lockbox agreement would allow rent payments to be made to the investor but would require timely deposit into the lockbox. A springing lockbox agreement calls for the establishment of a lockbox only when certain circumstances (e.g. borrower default) occur.

Fees at origination

It is not unusual for a lender to charge borrowers up-front fees. An **origination fee** typically charges a borrower a fixed percentage of the original loan balance. For example, a 1% origination fee on a $5,000,000 loan would require the borrower to pay the lender $50,000 when the loan is made. An **application fee** may also be charged by a lender. Application fees are often a fixed amount and not necessarily tied to the size of the loan requested by the investor.

54 Debt financing

Restrictions or fees relating to prepayment

There are a variety of reasons why a borrower may wish to pay off a loan in its entirety *before* the loan's original maturity date. For instance, an investor may wish to sell an investment property before the maturity date of the mortgage loan.[4] An investor may wish to refinance an existing loan before its maturity to take advantage of lower available interest rates in the market. If the market value of the investment property has risen significantly, an investor may wish to refinance before loan maturity to extract cash from a property without having to sell.

In general, lenders suffer when an investor prepays their mortgage loan. For instance, an investor refinancing to benefit from lower interest rates ends the original lender's ability to collect above-market interest. Further, as described later in this chapter, lenders invest heavily in the underwriting of the investment property before making a loan, and unexpected prepayment prevents lenders from recovering these costs over time through the monthly repayment. For these reasons, investment property lenders typically make it costly for investors to prepay. A mortgage loan may contain a **lockout provision**. The lockout provision prohibits prepayment. For example, a loan might contain a lockout provision for the first 2 or 3 years of a mortgage loan. An investment property mortgage may come with **prepayment fees**. These fees might specify a penalty equal to a percentage of the prepaid principal balance of the loan. Sometimes this penalty fee is constant throughout the life of the loan and sometimes the prepayment fee declines with the age of the loan. For example, one common prepayment fee is the 5-4-3-2-1, which means that a borrower prepaying in the first year faces a fee equal to 5% of the prepaid balance, a borrower prepaying in the second year faces a 4% fee, etc.

Yield maintenance clauses are designed to serve as a strong deterrent to loan prepayment. The name yield maintenance refers to the fact that a loan with such a clause will charge a prepaying borrower a fee that ensures that a lender, investing the fee at current risk-free interest rates, will be able to recreate the originally promised payments and therefore earn the same yield as it would have had the borrower not prepaid. The yield maintenance penalty can be calculated as the difference between (1) the present value of the remaining payments had the loan not been prepaid, where the discount rate is chosen to equal the Treasury rate at the maturity equal to the time that was originally remaining, and (2) the amount being prepaid. Both because loan prepayment tends to occur following interest rate declines and because Treasury rates are generally lower than mortgage rates, the value calculated in (1) will be larger than that calculated in (2), leading to a penalty. Note that with yield maintenance, not only is the lender getting the present value of all of the originally promised cash flows, it is also gaining the increased safety of receiving these cash flows on a Treasury-equivalent basis.

> Example 3.2: Consider the example loan described in Table 3.4. Calculate the yield maintenance penalty if you were to refinance this loan after making the first 7 years of payments.

Table 3.4 Refinanced loan

Loan Characteristics	
Original balance	$1,000,000
Interest rate	5.50% (APR)
Amortization	25 years
Maturity	10 years
Payments completed	84
Current yield on 3-year Treasuries	3.00%

Debt financing

To calculate the yield maintenance penalty, we proceed in several steps. First, we can determine that the monthly payment for this loan is $6140.87 = PMT(5.50% ÷ 12, 300, 1000000). Second, we need to calculate what the outstanding balance would be if the loan were carried to its original maturity. As mentioned previously, this can be calculated using an amortization table. Alternatively (and the way we shall proceed here), one can calculate the amount outstanding on an amortizing loan by calculating the present value of the remaining payments over the amortization period, discounted at the loan interest rate. After 10 years, there would be 15 years (180 months) of time left in the amortization period. Therefore, the originally scheduled repayment upon maturity can be calculated as $751,560.31 = PV(5.50% ÷ 12, 180, 6140.87). Third, we can calculate the present value of the remaining promised payments (e.g. 36 monthly payments of $6140.87 plus a final payment of $751,560.31), only now discounting at the *Treasury* rate. This is equal to the sum of the present value of the 36 monthly payments being prepaid, which we calculate as $211,162.98 = PV(3.00% ÷ 12, 36, 6140.87) and the present value of the originally scheduled final repayment, calculated as $686,951.55 = $751,560.31 ÷ (1 + 3.00% ÷ 12)^{36}. The sum of these two amounts is $898,114.53, which is the amount (1) described earlier. Note that with this cash flow, the lender could invest at the Treasury rate and replicate the originally scheduled cash flow owed on the loan, thereby preserving its original yield.[5] We next calculate how much we still owe on the mortgage at the time of prepayment. After 84 payments, the loan balance is $840,851.35 = PV(5.50% ÷ 12, 216, 6140.87). This is the amount (2) described above. The yield maintenance penalty is therefore $57,263.18, the difference between (1) and (2). This equates to 6.81% of the outstanding loan balance.

By design, the magnitude of the fee (relative to outstanding balance) will be higher if the original loan interest rate is higher or if the prevailing Treasury rate is lower. This is illustrated in Table 3.5, which illustrates the size of the yield maintenance fee in this example as one changes the assumption regarding the loan's original mortgage interest rate and the prevailing 3-year Treasury rate.

Thus, yield maintenance requirements act as a strong deterrent to the prepayment of a mortgage loan. Because of this, it is often the case that loans with yield maintenance carry lower interest rates than those with straight percentage prepayment fees. Often times, too, such loans are assumable and therefore the penalties may not be explicitly paid, but rather might be part of the negotiation between buyer and seller as part of a property sale transaction.

Like yield maintenance clauses, **defeasance** requirements also serve as a strong deterrent to mortgage loan prepayment. Mortgages with defeasance requirements do not technically allow the prepayment of a mortgage loan, but instead allow for the release of the lien on the collateral property from the existing mortgage so that the property investor can seek new

Table 3.5 Yield maintenance fee as a function of mortgage and Treasury rates

		\multicolumn{5}{c}{Mortgage rate}			
		4.50%	5.00%	5.50%	6.00%	6.50%
	2.00%	6.88%	8.27%	9.68%	11.09%	12.50%
	2.50%	5.46%	6.84%	8.23%	9.63%	11.03%
Treasury rate	3.00%	4.07%	5.43%	**6.81%**	8.19%	9.58%
	3.50%	2.69%	4.05%	5.41%	6.78%	8.15%
	4.00%	1.34%	2.68%	4.03%	5.38%	6.74%

56 Debt financing

financing. To have the original lien released, the investor must supply the lender with substitute collateral of higher quality – in practice, US Treasury securities – in an amount that is sufficient to deliver cash flow to the lender equal to what the lender had been promised in the original mortgage. So how much must a borrower pay to defease a loan? The answer is the present value of all remaining promised payments, discounted at the current Treasury rate specific to each payment. If the yield curve is flat, then the cost of buying replacement securities is equal to the cost of yield maintenance. With an upward sloping yield curve, defeasance will be more expensive than yield maintenance. As a practical matter, defeasance is typically facilitated by third parties who will charge a borrower a flat fee in addition to the cost of the replacement securities.

> Example 3.3: Consider the same loan described in Table 3.4. Let's demonstrate that the defeasance penalty is the same as the yield maintenance penalty if the Treasury yield curve was flat at 3% and you were not charged a defeasance fee.

For defeasance, you are required to buy Treasury securities that upon maturity will pay each remaining promised payment on the original loan. This is precisely equal to the present value of the remaining promised payments (e.g. 36 monthly payments of $6140.87 plus a final payment of $751,560.31), discounted at the *Treasury* rate specific to the timing of the payment. Because the yield curve is flat at 3%, this calculation is identical to the yield maintenance penalty. Of course, in practice, a borrower might be charged a fee of $50,000 for arranging defeasance, which typically makes defeasance more expensive than yield maintenance. With an upward sloping yield curve, the early payments are discounted at lower rates, which also makes the cost of defeasance rise above the cost of yield maintenance.

Lender underwriting of an investment property

Earlier in the chapter, we took loan terms as given and asked what influence those terms have on the debt service cash flows of a borrower. We now turn to how lenders consider which terms to offer. The process of evaluating the risks of lending to a property investor and establishing loan terms is called **underwriting**.

Motivation

As an investor, you want to invest in a property if it generates an expected return higher than (or at least not less than) is necessary to compensate you for the investment's systematic risk. To understand the terms of the loan we might get as a property investor, it is helpful to put yourself temporarily into the shoes of the lender. When a lender is contemplating providing mortgage finance to support your acquisition of a property, it faces an investment problem analogous to that of the investor, only now, the lender is contemplating investing in the debt that it provides you. The lender wants to make a loan that earns the lender a sufficient expected return.

How might the lender think about its expected return? First, define a lender's **yield** as the rate of return that the lender would earn if you, the investor, make all of the promised payments in full and on time. In practice, a lender's yield will be higher than the loan's interest rate due to the presence of a variety of fees associated with the loan. Note, however, that the yield is not the same as the expected return on the loan. In particular, the expected return on the loan will be lower than its yield because borrowers don't always make every promised payment on time, and in many instances borrowers will default on the loan and the lender will recover substantially less than was promised.

> Example 3.4: Suppose a bank lends $10,000,000 to an investor that uses the proceeds to partially finance the acquisition of a $15,000,000 investment property. The terms of the loan are very simple, requiring the borrower to pay $1,000,000 in interest each year for 3 years. At the end of the third year, the loan will mature and the mortgage contract requires the borrower to repay the $10,000,000. What is the lender's yield?

In the absence of any fees, the yield on this loan is 10%. This can be calculated as the internal rate of return (IRR) of the promised cash flows. In Year 0, the lender "invests" $10,000,000, it is promised $1,000,000 in Years 1 and 2, and is promised $11,000,000 in Year 3. The IRR of this payment stream, which by definition is the yield on the loan, is 10%, the interest rate on the loan.

A lender knows that sometimes its borrowers do not make all of their promised payments. What this implies is that yields do not always equal returns.

> Example 3.5: Suppose that the property loses a major tenant unexpectedly and the local market suffers a serious downturn. In combination, these factors lower the property's value to the point that the investor cannot repay the lender the $11,000,000 that is owed in the third year. As a result, the lender forecloses on the property, sells it in the market, and recovers a total of $7,700,000. What actual return did the lender realize?

The lender's realized return in this case can be calculated as -1.12%, which is the IRR calculated from the cash flow stream -10, 1, 1, 7.7.

> Example 3.6: Imagine that the lender believes that 90% of the time, the investor will be able to repay the loan as promised. However, 10% of the time, the borrower defaults as in Example 3.5. What is the lender's expected return?

In this example, the expected return to the lender is simply 8.89%, calculated as (10% × 90%) + (−1.12% × 10%) = 8.89%.

The mortgage loan in Example 3.6 has a 10% **probability of default (PD)**. This means that 10% of the time, the borrower will not make all of its payments in full and on time. The loan also had a 30% **loss given default (LGD)**. This reflects that the lender was promised to receive $11,000,000 in Year 3, but conditional on the borrower defaulting in Year 3, the lender only recovered $7,700,000. So, the lender's loss was $3,300,000 out of $11,000,000, or 30%.

Ideally lenders decide upon loan terms that provide it with an expected return appropriate for the risks of the loan. As illustrated by this simple example, a lender will choose loan terms to deliver a yield such that in combination with its views on PDs and LGDs it will earn an expected return warranted by the loan's risks. Ultimately, the required expected return on a mortgage loan should be driven by capital market factors.

Having now discussed why lenders wish to underwrite the loans they make, we now turn to the methods lenders use to set the terms of a loan such that their expected return compensates them for the risks they face.

Components of underwriting investment property

There are 3 major factors lenders consider when underwriting the risks of an investment property mortgage.

The property

The collateral property is the main source of repayment for the loan, so a lender focuses primarily on the question of whether or not the property will be able to generate adequate cash flow relative to the anticipated debt. Lenders have identified 3 metrics that they use to reflect the property's ability to support the loan. The first is the **debt service coverage ratio (DSCR)**. The debt service coverage ratio is defined as the property's annual net operating income (NOI) divided by the annual debt service (both principal and interest). A typical minimum standard for the DSCR might be something close to 1.25, although stabilized properties will typically have a ratio noticeably higher than this. Note that a DSCR equal to 1 means that every dollar of NOI is owed to the lender. Recall that since NOI is calculated before capital expenditures, a DSCR standard of 1.25 or higher implies that additional cash flow exists after debt service to fund these additional capital expenditures.

One subtlety in the calculation of the DSCR is that lenders will typically begin with a pro forma constructed by the investor. However, they will adjust the figures according to their own standards so that the NOI they estimate is one that they are comfortable lending against. These adjustments will vary across lenders, but might include adjusting rents and occupancy to those typical of the market, distinguishing base rent from reimbursements and concessions to get a more consistent picture of rental income, adjusting management expenses to market levels to undo the overestimation of income for self- or affiliate-managed property, adjusting property tax levels if a reassessment is likely, and adjusting TIs and other capital costs to reflect typical levels and the perceived risks of tenant rollover. This lender-adjusted measure of NOI is sometimes called underwritten NOI.

The DSCR is a metric used by lenders to give them comfort that the investor will be able to meet ongoing debt service during the holding period. However, having a high DSCR does not guarantee an investor's ability to repay the loan at maturity. As we saw earlier in this chapter, loans against investment property are not typically fully amortizing, and therefore lenders face a risk that investors will not be able to pay the balloon payment at maturity. Typically, the loan payoff required at loan maturity is paid from the proceeds of a new loan, so a lender wanting to have confidence that its loan will be repaid in full is generally looking to ensure that the investor will be able to refinance in the future.

The metric commonly used to measure a property's ability to get future financing is the **loan-to-value ratio (LTV)**. As its name would imply, a loan's LTV is the ratio of the size of the outstanding loan balance divided by the value of the property. The terms of the loan will dictate the outstanding loan balance – both at loan origination and at loan maturity. However, the denominator in the LTV might be valued by a lender in a variety of ways. It may use a DCF approach after making underwriting adjustments to the investor's pro forma. It may look to comparable sales, typically by looking at recent transactions to get an estimate of cap rates for similar properties. It may also be guided by measures of replacement cost. Overall, it is important for an investor to understand that a lender need not share the same opinion about the value of the collateral property. A typical LTV requirement for an investment property mortgage is near 60%, although loans secured by apartments are typically higher, near 70%. Lenders require an LTV less than 100% to ensure that investors have an equity stake in the property. This equity stake incentivizes the property owner to effectively manage the property and to make the monthly repayments. Lower LTVs also provide lenders with greater protection against property price declines. That is, a loan with a lower initial LTV will be more likely to be refinanced upon maturity than a loan on the same property with a higher initial LTV. Thus, lenders view loans with lower LTVs as less risky.

A final metric that has become increasingly used by lenders in recent years is the **debt yield**. What first confuses investors is that the debt yield, despite its name, is not the yield on the debt from the perspective of the lender. The underwriting metric debt yield is defined as the property's (underwritten) NOI divided by the initial mortgage balance of the loan. The intuition for lenders' use of this metric is that the debt yield is a measure of the lender's income return following an immediate foreclosure. In other words, the debt yield measures the implied cap rate of the property holding income constant but lowering the value of the property to the outstanding balance of the loan.

The credit and experience of the borrower

The property is the primary source of loan repayment, which is why lenders focus primarily on the ability of the property to generate cash flow to pay debt service and to be refinanced at loan maturity. However, lenders will also consider the credit and experience of the borrower.[6] Even when the loan is non-recourse, the lender will examine whether the borrower has experience in property ownership and operation. Further, the lender will check whether the investor has previously made timely payments to lenders, been responsive to lender inquiries, and been cooperative or litigious when things have gone wrong. Finally, the lender might be interested in the borrower's other current property investments. If these other investments are struggling, a lender might worry that the borrower would divert cash flow from the new property to pay off other debt.

The market in which the property is located

The location of the property provides the lender with context for the ultimate performance of the investment property. For example, if the collateral is retail property, a lender would want to see that the location of the building is in a high-traffic area. If the collateral is a hotel catering to business travelers, the lender might expect it to be located near offices with large corporate tenants. More generally, the lender is interested in the underlying economic activity of the area. For example, what is the employment growth in the local area? Is the economic base of the community well diversified? Are there important demographic trends that could negatively influence the performance of the property going forward? Will new properties that could compete with the existing property be constructed soon? Are there barriers to new development? Is the supply and demand for space currently in a state of balance?

Loan structuring

When an investor applies for a loan to support the purchase of an investment property, the loan ultimately offered by a lender will reflect the underwriting process that we have just described. It is important that investors realize that there is more to a loan than its quoted current interest rate. The level of the rate is largely influenced by the riskiness of the property (and borrower), the level of interest rates in the capital markets, and the ultimate structure of the loan – that is, the other features of the loan that the lender determines.

For instance, how much an investor will be able to borrow is driven by a lender's LTV, DSCR, and debt yield constraints. A lender can choose an amortization period to affect trade-offs between these constraints. For instance, a longer amortization period will reduce debt service and therefore increase the loan's DSCR. However, the longer amortization period will increase the balloon payment due upon loan maturity, increasing refinancing risk and

60 Debt financing

therefore likely causing the lender to insist upon a lower LTV. Lenders may also "split" loans into pieces. For instance, a lender may not like the risk of lending 70% LTV against an investment property even if it could charge a relatively high interest rate of 7.5%. It might, however, be perfectly happy to lend 60% LTV at 6% and another (or the same) lender might be willing to lend an additional 10% LTV (subordinated to the first loan) at 12%. This may benefit the borrower since $\frac{6}{7} \times 6\% + \frac{1}{7} \times 12\% < 7.5\%$. Finally, lenders might impose additional restrictions on investors before being willing to provide a mortgage. One common restriction is that certain reserves must be set aside by the investor to address specific capital needs of the property that were identified during the underwriting process.

Underwriting illustrations

Example 3.7: Suppose an investor owns a property that all lenders agree is worth $2,500,000 and has an underwritten level of NOI of $187,500. Furthermore, suppose all lenders set the mortgage interest rate at 5% and the amortization period at 25 years, although these values, too, would be part of a more general underwriting process. How do variations in required underwriting metrics across lenders affect the amount an investor can borrow?

The top panel of Table 3.6 illustrates minimum underwriting criteria that 3 lenders have applied. Note that in each column, there are different DSCR, LTV, and debt yield constraints. The second panel of Table 3.6 illustrates the calculations needed to determine the maximum loan size the investor will be able to receive. Beginning with the loan offer from Lender A, a maximum mortgage size of $1,562,500.00 is possible. This mortgage generates $117,187.50 in debt service, which can be calculated as 12 × PMT(5% ÷ 12, 300, 1562500). This debt service generates a DSCR of 1.71 = $187,500 ÷ $117,187.50. The LTV on the loan would be 62.5% = $1,562,500 ÷ $2,500,000. The debt yield is 12% = $187,500 ÷ $2,500,000. In this example, the debt yield constraint is binding. That is, the loan size cannot be increased above $1,562,500 because doing so would reduce the debt yield below 12%, which is assumed to be the minimum that the lender will accept.

The underwriting standards used by Lender B and Lender C were chosen such that the binding constraint becomes the DSCR (column 2) or the LTV (column 3). In general, only 1 constraint will be binding. To see why, consider the calculations shown in the bottom panel of Table 3.6. The first row calculates the maximum annual debt service of a loan, conditional on the property NOI of $187,500 and the DSCR ratio required by the lender. For example, the value of $117,187.50 shown in the first column can be calculated as property NOI divided by the DSCR limit ($117,187.50 = $187,500 ÷ 1.6). Dividing that number by 12 gives the maximum allowable monthly debt service of $9765.63. From the monthly debt service, one can calculate the size of the largest loan that satisfies the DSCR by calculating the present value of 25 years (the amortization period) of monthly payments of $9765.63. This yields a loan value of $1,670,508.27 = PV(5% ÷ 12, 300, 9765.63). The interpretation of the value $1,670,508.27 is the maximum loan that is possible if the only constraint used by the lender was that the DSCR had to meet or exceed 1.6. The next row calculates the maximum loan size subject only to the LTV constraint. In column 1, the LTV constraint is given to be 65%, which implies that a $2,500,000 property can support a loan size of $1,625,000. The final row maximizes the loan size subject to a debt yield greater than or equal to 12%. This can be calculated directly as $1,562,500 = $187,500 ÷ 12%. Thus, each of the 3 constraints provides a bound on the value of the mortgage size. Only by choosing the minimum of these 3 mortgage sizes will all 3 constraints be met simultaneously. Thus, the

Table 3.6 How underwriting metrics affect loan size

	Lender A	Lender B	Lender C
Property value	$2,500,000.00	$2,500,000.00	$2,500,000.00
Underwritten NOI	$187,500.00	$187,500.00	$187,500.00
Required DSCR	1.6	1.8	1.4
Required LTV	65.00%	70.00%	60.00%
Required debt yield	12.00%	10.00%	11.00%
Interest rate	5%	5%	5%
Amortization Period	25	25	25
Mortgage amount	**$1,562,500.00**	**$1,484,896.24**	**$1,500,000.00**
Implied debt service	$109,610.63	$104,166.67	$105,226.21
Implied DSCR	1.71	**1.80**	1.78
Implied LTV	62.50%	59.40%	**60.00%**
Implied debt yield	**12.00%**	12.63%	12.50%
Maximum annual debt service	$117,187.50	$104,166.67	$133,928.57
Maximum monthly debt service	$9765.63	$8680.56	$11,160.71
Maximum loan size from DSCR	$1,670,508.27	$1,484,896.24	$1,909,152.31
Maximum loan size from LTV	$1,625,000.00	$1,750,000.00	$1,500,000.00
Maximum loan size from debt yield	$1,562,500.00	$1,875,000.00	$1,704,545.45

maximum mortgage shown in the second panel of Table 3.6 is derived from the minimum of the 3 separate calculations shown in the bottom panel.

Sources of debt finance for investment property

We conclude Chapter 3 by discussing some nuances in mortgage lending that relate to the type of institution originating the mortgage loan. In particular, different types of lending institutions typically offer credit on different terms.

Balance sheet lenders

A **balance sheet lender** is an institution that internalizes all the components of mortgage lending. That is, a balance sheet lender will originate loans, service loans (i.e. collect monthly payments), and renegotiate loans to borrowers that fall into distress. The most well-known and common type of balance sheet lender is a commercial bank. Well over half of all the debt financing for investment property is originated by these traditional lenders. The typical commercial bank funds itself largely through short- to medium-term deposits. This leads such lenders to typically offer commercial mortgages with similarly short- to medium-term maturities. Another common balance sheet lender is a life insurance company. These insurers tend to have noticeably longer-term liabilities than commercial banks. Therefore, life insurers are typically willing to lend against investment property for longer periods of time. However, because they are interested in lending over longer horizons, life insurance companies tend to make safer loans. Not only do life insurance companies lend against higher quality properties in safer markets, but they also tend to make loans with lower LTVs relative to commercial bank mortgages. Finally, there are balance sheet lenders that are organized as private equity funds, mortgage Real Estate Investment Trusts (mortgage REITs), or hedge funds. These types of institutions typically emphasize speed and flexibility relative to the more traditional

62 Debt financing

lenders. They also typically seek higher rates of return. From the borrower's perspective, therefore, the advantage of greater loan flexibility comes at the cost of a higher interest rate.

Conduit lenders

A **conduit lender** is an institution that underwrites and originates mortgages backed by investment property but sells the loans soon after origination in a process known as securitization. Thus, the implicit lenders in the securitization are bond investors buying bonds whose cash flows are backed by the interest and principal payments of the actual mortgage loans. The process is rather complicated, but the key distinguishing characteristic of conduit lending is that the terms offered by a conduit lender are implicitly determined by the market for mortgage-backed bonds. Thus, the terms offered by conduit lenders tend to be more volatile than those of balance sheet lenders.

Does the source of debt matter to the property investor?

As this chapter has indicated, there are many features of a mortgage other than its interest rate. Often times, a property investor will approach multiple lenders or a mortgage broker that represents multiple lenders to determine which financing terms are available for a given property. The different types of lenders will generally offer different terms. For instance, defeasance requirements are common in conduit lending because bond investors want certainty in the cash flows they will receive. More generally, there will typically not be one lender that offers the best terms in every dimension. Therefore, a property investor deciding between loan offers will have to consider which loan features are most important *for the given loan.*

One potentially important feature of a mortgage loan that may not be readily apparent at origination is the lender's flexibility should the property investor fall into distress. Although we will discuss what might happen when a property investor faces distress in greater detail in Chapter 7, it is worth mentioning that a balance sheet lender would be able to renegotiate the loan in any way it deemed necessary to benefit its position as a creditor of the investor. By contrast, a conduit loan is serviced by a specialist hired to work in the interest of the bondholders. Thus, loans that have been originated by conduit lenders and securitized typically have less ex-post renegotiation possibilities. Investors seeking loans on riskier properties might value renegotiation flexibility. However, since banks and insurance companies tend to prefer safer borrowers and higher quality collateral, loans backed by risky properties are typically financed by conduit lenders.

Debt financing: a tale of 2 properties

In this case, you are put in the shoes of Stanley Cirano, owner of 2 suburban Chicago retail centers. With interest rates lower than they've been in some time, Cirano considers whether a refinancing of one or both of his properties is worthwhile. After reading and analyzing the case, you will be able to:

1 Evaluate debt offerings from different lenders and understand their strengths and weaknesses.
2 Calculate the cost of refinancing and understand the many non-price components of a commercial mortgage.
3 Evaluate a lending opportunity from the lender's perspective, especially on a property with uncertain prospects.

A tale of 2 properties: debt strategies for financing commercial real estate

With interest rates near an all-time low in late 2015, Stanley Cirano decided to review the financing on his holdings of commercial real estate. Cirano Properties was the owner of 2 separate retail shopping centers in suburban Chicago.

The first was Brookline Road Shopping Center, which Cirano had acquired in 2006 and had managed through the financial crisis and real estate downturn. The property was performing well, and Cirano wondered whether he should refinance or sell it. The second property was Columbus Festival Plaza, which Cirano had acquired in a bankruptcy auction in 2010. Although the property had required a large amount of capital improvement, Cirano was proud of the growth of NOI he had generated.

The 2 properties had many similarities, including that they were both significantly financed with debt. At the time he purchased the properties, Cirano simply borrowed as much as he could from the lender with the lowest interest rate he could find. Since that time, Cirano had learned that there was more to the optimal financing of commercial property than just dollars and rates. Nonetheless, Cirano was convinced that interest rates would soon rise, so he thought it made sense to consider the debt options available to him and make a sound, strategic decision on the financing of both of his assets at the same time.

Stanley Cirano

Born and raised on the north side of Chicago, Stanley Cirano had been attracted – as many people were – by the weather and opportunities in California's technology sector in the mid-1990s. A bit of good timing landed Cirano a job at an internet startup about 18 months before it went public. Shortly before the dot-com crash, Cirano sold his shares for over $8,000,000.

Cirano's windfall attracted the attention of private wealth managers, who convinced him to invest in commercial real estate. To try it out, in 2001 Cirano invested $500,000 in a triple net lease property occupied by a branch of the US Postal Service. The investment was a perfect introduction to commercial real estate – it required a relatively small equity investment and carried little risk because the government's lease ran through 2011. When Cirano sold the property in 2005, he acquired a little more wealth and, more importantly, much more confidence in his ability to make a career in commercial real estate. Thus, Cirano Properties was born.

Brookline road shopping center

After his initial success in real estate, Cirano sought to increase his investments in the sector. He invested $3,500,000 of equity towards the $10,000,000 acquisition of Brookline Road Shopping Center. The 65,611-square-foot shopping center was located in Buffalo Lakes, a suburb northwest of Chicago with a median household income of $128,876. The property sat at the high-visibility corner of Brookline and Mundee Roads.

Brookline Road Shopping Center was built in 2004, and Cirano acquired the property from the developer 2 years later. Cirano liked the new construction and excellent location, and the property enabled him to rent the space that the developer had not yet preleased. Cirano successfully grew Brookline's NOI from $754,000 in 2006 to nearly $900,000 in 2015 (see Table 3.7). Cirano believed that he had navigated the financial crisis well, and the 2010 leasing of the final vacant space to Chipotle was a major factor in the property's income growth (see Table 3.8).

Table 3.7 Historical performance of Brookline Road Shopping Center

	2006	2007	2008	2009	2010	2011	2012	2013	2014	2015
Net rental revenue	$1,355,524.50	$1,361,034.50	$1,366,654.70	$1,242,303.00	$1,458,113.00	$1,517,764.20	$1,527,159.48	$1,536,742.67	$1,546,517.53	$1,562,825.38
Operating expenses										
Landscaping/snow removal	$46,000.00	$46,000.00	$46,000.00	$46,000.00	$46,000.00	$48,300.00	$48,300.00	$48,300.00	$48,300.00	$48,300.00
Cleaning/common area exp	$169,000.00	$169,000.00	$169,000.00	$169,000.00	$169,000.00	$175,000.00	$175,000.00	$175,000.00	$175,000.00	$175,000.00
Repairs and maintenance	$52,000.00	$53,040.00	$53,900.00	$56,200.00	$57,300.00	$58,500.00	$59,600.00	$60,900.00	$61,900.00	$63,400.00
Utilities	$153,000.00	$156,060.00	$159,200.00	$162,400.00	$165,600.00	$169,100.00	$172,800.00	$176,100.00	$180,200.00	$184,200.00
Real estate taxes	$129,000.00	$129,000.00	$129,000.00	$129,000.00	$129,000.00	$136,000.00	$136,000.00	$136,000.00	$136,000.00	$136,000.00
Management fee	$40,665.74	$40,831.04	$40,999.64	$37,269.09	$43,743.39	$45,532.93	$45,814.78	$46,102.28	$46,395.53	$46,884.76
Insurance	$12,000.00	$12,204.00	$12,411.47	$12,622.46	$12,837.04	$13,055.27	$13,277.21	$13,502.93	$13,732.48	$13,965.93
Total operating expenses	$601,665.74	$606,135.04	$610,511.11	$612,491.55	$623,480.43	$645,488.20	$650,792.00	$655,905.21	$661,528.00	$667,750.69
Net operating income	$753,858.77	$754,899.47	$756,143.59	$629,811.45	$834,632.57	$872,276.00	$876,367.49	$880,837.47	$884,989.52	$895,074.69
Capital costs										
Tenant improvements	$0.00	$0.00	$0.00	$36,000.00	$0.00	$0.00	$0.00	$0.00	$0.00	$0.00
Leasing commissions	$0.00	$0.00	$0.00	$0.00	$0.00	$0.00	$0.00	$0.00	$0.00	$0.00
Capital reserves	$45,231.53	$45,293.97	$45,368.62	$37,788.69	$50,077.95	$52,336.56	$52,582.05	$52,850.25	$53,099.37	$53,704.48
Total capital costs	$45,231.53	$45,293.97	$45,368.62	$73,788.69	$50,077.95	$52,336.56	$52,582.05	$52,850.25	$53,099.37	$53,704.48
Operating cash flow	$708,627.24	$709,605.50	$710,774.98	$556,022.76	$784,554.61	$819,939.44	$823,785.44	$827,987.22	$831,890.15	$841,370.21
Debt service	$503,316.82	$503,316.82	$503,316.82	$503,316.82	$503,316.82	$503,316.82	$503,316.82	$503,316.82	$503,316.82	$503,316.82
DSCR	1.50	1.50	1.50	1.25	1.66	1.73	1.74	1.75	1.76	1.78
Post-debt service proceeds	$205,310.42	$206,288.67	$207,458.15	$52,705.94	$281,237.79	$316,622.62	$320,468.61	$324,670.40	$328,573.33	$338,053.38

Tenant	Sq. Ft.	2006	2007	2008	2009	2010	2011	2012	2013	2014	2015
Jewel-Osco Grocery Store	35,000										
Rate		$19.25	$19.25	$19.25	$19.25	$19.25	$20.50	$20.50	$20.50	$20.50	$20.50
Income		$673,750.00	$673,750.00	$673,750.00	$673,750.00	$673,750.00	$717,500.00	$717,500.00	$717,500.00	$717,500.00	$717,500.00
Wells Fargo Bank	9,500										
Rate		$29.00	$29.58	$30.17	$24.00	$24.48	$24.97	$25.47	$25.98	$26.50	$27.03
Income		$275,500.00	$281,010.00	$286,630.20	$228,000.00	$232,560.00	$237,211.20	$241,955.42	$246,794.53	$251,730.42	$256,765.03
Chipotle	6,500										
Rate		$0.00	$0.00	$0.00	$0.00	$32.50	$32.50	$32.50	$32.50	$32.50	$33.48
Income		$0.00	$0.00	$0.00	$0.00	$211,250.00	$211,250.00	$211,250.00	$211,250.00	$211,250.00	$217,587.50
General Dentistry Office	4,500										
Rate		$24.00	$24.00	$24.00	$24.00	$24.00	$26.50	$26.50	$26.50	$26.50	$26.50
Income		$108,000.00	$108,000.00	$108,000.00	$108,000.00	$108,000.00	$119,250.00	$119,250.00	$119,250.00	$119,250.00	$119,250.00
FedEx Ship Center	10,111										
Rate		$29.50	$29.50	$29.50	$23.00	$23.00	$23.00	$23.46	$23.93	$24.41	$24.90
Income		$298,274.50	$298,274.50	$298,274.50	$232,553.00	$232,553.00	$232,553.00	$237,204.06	$241,948.14	$246,787.10	$251,722.85
Total	65,611	$1,355,524.50	$1,361,034.50	$1,366,654.70	$1,242,303.00	$1,458,113.00	$1,517,764.20	$1,527,159.48	$1,536,742.67	$1,546,517.53	$1,562,825.38

LEASE INFORMATION

Tenant	2015 Rent ($ per sq. ft.)	Expiration	Additional Notes
Jewel-Osco Grocery Store	20.50	2025	Signed lease extension (through 2025) in 2011. Rent will increase to $23.50 per square foot in 2020.
Wells Fargo Bank	27.03	2024	Signed lease extension (through 2024) in 2009. Rent was dropped to retain tenant during financial crisis.
Chipotle	33.48	2024	Signed 15-year lease in 2010. Rent flat for 5 years, then grows 3% per year until end of term.
General Dentistry Office	26.50	2021	In the middle of long-term lease. Rate increased to $26.50 per square foot in 2011, adjusting to $29.00 in 2016, where it remains until the end of the lease. Tenant has full set of patients from the community.
FedEx Ship Center	24.90	2019	Dropped rent in 2009 in order to re-sign tenant to the space. Rent flat for 3 years, then grows 2% per year until end of term.

66 Debt financing

For the past few months, Cirano had wondered whether now was the time to sell. Cirano knew some real estate investors were cashing out, but others wanted to wait until they found an attractive opportunity in which to invest any sales proceeds. Alternatively, perhaps a refinancing would allow him to recoup much of his initial investment while continuing to receive the steady and growing cash flows from the property.

The original $6,500,000 loan on the property had been originated by Skyline Bank. The loan had 3 months remaining on its original 10-year maturity. It had been amortizing on a 30-year schedule and carried an interest rate of 6.70%. The loan's 5-year lockout provision[7] had expired in 2011, and a prepayment penalty of 3% of total prepaid balances was waived within the final 6 months before maturity.

Over the past 10 years, Cirano had deepened his relationship with Skyline Bank. The bank held his business and personal accounts, and he also considered many of the bank's managers his personal friends, with whom he enjoyed an occasional round of golf. During a recent golf outing with Cirano, Skyline vice president Charlie Darvin outlined what the bank would be willing to provide to Cirano for ongoing debt financing of Brookline. For a total of $20,000 in fees, Skyline would offer a loan of up to 65% LTV as long as the property also had a minimum DSCR of 1.25.[8] The loan would have a new maturity of 10 years, amortize over 30 years, and carry an interest rate of 4.375%. The new loan would still contain Skyline's standard methods of discouraging prepayment – a 5-year lockout provision followed by a 3% prepayment penalty. Like its original loan, Skyline's offering required full recourse to Cirano in the event of default.

However, Skyline was not the only lender interested in refinancing Brookline. Adam Jones, a loan officer with regional lender Fourth Second Bank, had been courting Cirano. In exchange for a 2% origination fee, Fourth Second would lend up to 60% LTV with an interest rate of 3.625% as long as the property's DSCR was at least 1.30. The loan would mature in 15 years and amortize over 25. The loan was prepayable without cost after an initial 2-year lockout. It, too, carried full recourse to Cirano.

Finally, Cirano heard from his friend Raymond Ross, a mortgage broker in Chicago, that Ross's client Regional Liberty Life Insurance Company was looking to place a large amount of commercial real estate debt in the Chicagoland market. According to Ross, Regional Liberty would refinance Brookline on a non-recourse basis up to 70% LTV at an interest rate of 4.625% as long as the property's DSCR was above 1.20. The loan had a 1.5% origination fee, a 20-year maturity, and a 30-year amortization. Regional Liberty loans carried a 5-year lockout on prepayment, followed by a 5-4-3-2-1 prepayment penalty in Years 6–10.[9] Ross explained that Cirano would have little direct contact with Regional Liberty after the due diligence process.

Cirano made some notes of his financing choices for Brookline (see Table 3.9).

Table 3.9 Brookline Road Shopping Center debt financing choices

	Skyline Bank	Fourth Second	Regional Liberty
Origination fees	$20,000	2%	1.50%
Interest rate	4.375%	3.625%	4.625%
Maximum LTV	65%	60%	70%
Minimum DSCR	1.25	1.30	1.20
Maturity	10	15	20
Amortization	30	25	30
Prepayment penalties	5-year lockout; 3% afterward	2-year lockout	5-year lockout; 5-4-3-2-1 afterward
Recourse	Yes	Yes	No

Columbus Festival Plaza

While managing Brookline, Cirano had contemplated whether to pursue other investment opportunities. Cirano had not been actively looking for commercial property in late 2009 and early 2010 because he was focused on renegotiating leases at Brookline and keeping that property above water during the severe downturn. However, Cirano heard through his network about a bankruptcy auction for a number of nearby retail centers, and when he saw the low minimum bids he thought it was worth the effort to investigate. This is how Cirano ultimately acquired Columbus Festival Plaza, an 113,572-square-foot Class B shopping center in Elm Park, a western suburb of Chicago with a median household income of $63,537.

Cirano bid $8,700,000 for Columbus – a low bid with a long shot to win. As it turned out, however, investors were only slowly getting back into suburban commercial real estate markets, and his bid was sufficient. The property, which had been built in 1974 and remodeled in 1992, was in a favorable location at the corner of Columbus and Crawford and had high traffic visibility. Two other retail strip centers shared the same intersection. Columbus had some vacant space that Cirano thought would be filled once the economy improved.

Debt markets in 2010 were still rather tight, so Cirano financed the Columbus acquisition with 45% equity and 55% debt. The non-recourse, nearly $4,800,000 mortgage was originated by Principal Property Finance at a rate of 5.9%, a 9-year maturity, and with amortization over 30 years. Principal immediately sold the loan into a securitization. The loan contained a defeasance clause, which meant that Cirano could not simply pay off the debt; he would have to replace the collateral on the loan with sufficient Treasury securities that would cover all the remaining payments due on the note. In retrospect, he realized he should have paid closer attention to the note's legal language, but this was his first and only securitized debt, and he simply had missed it. Defeasance was a bit tricky, but Cirano knew that there were defeasance specialists that would facilitate the defeasance for a fee of $50,000. That amount, plus the cost of the replacement securities, would be the total cost to defease the loan. He had made the first 72 payments on the mortgage, so there were only 36 more payments needed to defease.

Largely driven by declining vacancy since taking over the property, Cirano grew NOI from $765,000 in 2010 to just over $1,000,000 in 2015 (see Table 3.10). The most exciting development for the property was CVS Pharmacy's signing of a long-term lease on a newly built outlot on the property (see Table 3.11). The construction project was approved in 2014 and the store was expected to open in early 2016. Cirano did not need any additional equity to finance the outlot construction because the $1,750,000 project was entirely financed by a construction loan at 6%, which would come due upon CVS's 2016 occupancy.[10] When CVS occupied its new space, Cirano expected operating costs to increase by 11%.

Cirano could continue to hold the property until his current loan matured, allowing him to avoid the defeasance cost. However, Cirano wondered whether he should sell immediately, cashing in quickly on the CVS deal. Alternatively, he could refinance into a loan with a significantly lower interest rate. Ideally, refinancing would generate sufficient proceeds to pay off the maturing construction loan as well as defease the original debt. He first approached Skyline Bank, but Darvin told him that Skyline had high underwriting standards for tenant quality and that it would not be able to lend against Columbus. He then called Ross, who shopped the deal to his potential lenders.

Table 3.10 Historical financial performance of Columbus Festival Plaza

	2010	2011	2012	2013	2014	2015
Rental revenue	$1,757,224.00	$1,789,434.00	$1,791,224.00	$2,050,317.00	$2,058,333.06	$2,143,907.64
Operating expenses						
Landscaping/snow removal	$86,000.00	$86,000.00	$86,000.00	$86,000.00	$86,000.00	$93,000.00
Cleaning/common area exp	$211,000.00	$211,000.00	$211,000.00	$211,000.00	$211,000.00	$235,000.00
Repairs and maintenance	$123,000.00	$126,700.00	$133,400.00	$138,900.00	$141,600.00	$146,200.00
Utilities	$268,000.00	$273,900.00	$279,900.00	$287,100.00	$292,800.00	$299,400.00
Real estate taxes	$232,000.00	$232,000.00	$232,000.00	$232,000.00	$232,000.00	$249,000.00
Management fee	$52,716.72	$53,683.02	$53,736.72	$61,509.51	$61,749.99	$64,317.23
Insurance	$19,400.00	$19,788.00	$20,183.76	$20,587.44	$20,999.18	$21,419.17
Total operating expenses	$992,116.72	$1,003,071.02	$1,016,220.48	$1,037,096.95	$1,046,149.18	$1,108,336.40
Net operating income	$765,107.28	$786,362.98	$775,003.52	$1,013,220.05	$1,012,183.88	$1,035,571.24
Capital costs						
Tenant improvements	$117,250.00	$84,500.00	$0.00	$84,000.00	$0.00	$0.00
Leasing commissions	$0.00	$0.00	$0.00	$8505.00	$0.00	$0.00
Capital reserves	$68,859.66	$70,772.67	$69,750.32	$91,189.80	$91,096.55	$93,201.41
Total capital costs	$186,109.66	$155,272.67	$69,750.32	$183,694.80	$91,096.55	$93,201.41
Operating cash flow	$578,997.62	$631,090.31	$705,253.20	$829,525.25	$921,087.33	$942,369.83
Debt service	$340,578.98	$340,578.98	$340,578.98	$340,578.98	$340,578.98	$340,578.98
DSCR	2.25	2.31	2.28	2.97	2.97	3.04
Post-debt service proceeds	$238,418.64	$290,511.33	$364,674.22	$488,946.27	$580,508.35	$601,790.85

	Sq. Ft.	2010	2011	2012	2013	2014	2015
JC Penney	35,000						
Rate		$19.25	$19.25	$19.25	$19.25	$19.25	$20.50
Income		$673,750.00	$673,750.00	$673,750.00	$673,750.00	$673,750.00	$717,500.00
Valdez Value Groceries	22,456						
Rate		$16.50	$16.50	$16.50	$16.50	$16.50	$18.00
Income		$370,524.00	$370,524.00	$370,524.00	$370,524.00	$370,524.00	$404,208.00
Subway	4,500						
Rate		$24.00	$24.00	$24.00	$27.00	$27.00	$27.00
Income		$108,000.00	$108,000.00	$108,000.00	$121,500.00	$121,500.00	$121,500.00
SportRite Sporting Goods	14,555						
Rate		$20.00	$20.00	$20.00	$20.00	$20.00	$20.00
Income		$291,100.00	$291,100.00	$291,100.00	$291,100.00	$291,100.00	$291,100.00
Johnny B's Dry Cleaners	4,500						
Rate		$24.00	$24.00	$24.00	$26.50	$26.50	$26.50
Income		$108,000.00	$108,000.00	$108,000.00	$119,250.00	$119,250.00	$119,250.00
Italian Stallion Training Gym	8,950						
Rate		$23.00	$17.80	$18.00	$18.20	$18.40	$18.60
Income		$205,850.00	$159,310.00	$161,100.00	$162,890.00	$164,680.00	$166,470.00
Vicenzo's Pizza	4,500						
Rate			$17.50	$17.50	$17.50	$17.85	$18.21
Income			$78,750.00	$78,750.00	$78,750.00	$80,325.00	$81,931.50
First Bank Credit Union	10,111						
Rate					$23.00	$23.46	$23.93
Income					$232,553.00	$237,204.06	$241,948.14
CVS Pharmacy	13,500						
Rate							
Income							
Vacant Units	9,000						
Rate		$0.00	$0.00	$0.00	$0.00	$0.00	$0.00
Income		$0.00	$0.00	$0.00	$0.00	$0.00	$0.00
Total	113,572	$1,757,224.00	$1,789,434.00	$1,791,224.00	$2,050,317.00	$2,058,333.06	$2,143,907.64

LEASE INFORMATION

Tenant	Current Rent ($ per sq. ft.)	Expiration	Additional Notes
JC Penney	20.50	2017	Inherited lease. Rent flat at $20.50 per square foot until expiration. Retailer has been struggling, but sales are decent at this location.
Valdez Value Groceries	18.00	2020	Signed 10-year lease extension in 2010. Rent flat at $18.00 per square foot until expiration. Lower rent used to retain tenant during financial crisis.
Subway	27.00	2023	Inherited lease. Strong location for the franchise. Rent grows to $30.00 per square foot in 2018.

70 Debt financing

Table 3.12 Columbus Festival Plaza current debt and new financing choices

	Existing Loan	Oakwood Commercial Mortgage	Northwood National Bank
Origination fees	NA	1.10%	1.80%
Interest rate	5.900%	4.900%	4.125%
Original balance	$4,785,000.00	NA	NA
Maximum LTV	NA	65%	60%
Minimum DSCR	NA	1.25	1.40
Maturity	9	3	10
Amortization	30	25	30
Payments completed	72	NA	NA
Prepayment penalties	Defeasance	Defeasance	3-year lockout; 3% thereafter
Recourse	No	No	Yes

One letter of interest came from a conduit lender, a company that originated commercial mortgages and held them as investments for a short time before selling or securitizing the loans. The lender, Oakwood Commercial Mortgage, offered to lend on a non-recourse basis up to 65% LTV for a 3-year term with 25-year amortization at an interest rate of 4.9%. Like his current loan, Oakwood's loan also had a defeasance requirement. Oakwood charged a 1.1% origination fee and required a minimum DSCR of 1.25.

Another offer came from Northwood National Bank, a $1.6 billion bank headquartered in Des Moines, Iowa. Northwood planned to keep the loan on its balance sheet to diversify its portfolio of commercial real estate loans by adding exposure to markets outside – but not far from – its home state. Northwood expected a 1.8% origination fee but offered Cirano a 10-year loan with 30-year amortization at an interest rate of 4.125%. The loan had a maximum LTV of 60% and a minimum DSCR of 1.40. The loan locked out prepayment for 3 years and had a 3% prepayment penalty thereafter. Unlike the conduit offering, Northwood's loan would require full recourse to Cirano in the event of default.

Cirano wrote up his refinancing options for Columbus (see Table 3.12).

Next steps

Cirano had some thinking to do. Taking a holistic approach to the financing of his 2 property investments made sense to him, but even though interest rates were at an all-time low, his ability to obtain a good rate varied across his properties. He understood why that was the case but had not fully appreciated how much the characteristics of each property would affect his financing options.

His first task was to estimate how much debt each lender would be willing to provide. This was an exercise involving each lender's LTV and DSCR constraints. As a first step, he gathered data on comparable property sales around both of his investments (see Table 3.13) and for the Chicago retail market in general (Table 3.14). These data would help him estimate the value that prospective lenders would place on his properties. However, that was only a guideline and did not account for any features unique to his properties or any lender-specific underwriting considerations. For a better understanding, he outlined a list of assumptions (Table 3.15) that were necessary to understand his rent rolls going forward (Table 3.16 and 3.17). These assumptions allowed him to develop a cash flow pro forma for

Table 3.13 Comparable sales

Brookline Road Shopping Center

Date	Distance from Brookline (miles)	Property Name	Sq. Ft.	Year Built	Price ($)	$ per Sq. Ft.	Cap Rate	Occupancy	Comments/Notes
Apr-15	1.22	Copley Center	65,864	1989	10,600,000	161	7.2%	89%	Strip/unanchored property; tenants: The Room Place, Fruitful Yield, Salata, The Cosmetology and Spa Institute; prior sale: Sept. 1999 ($6.8 mil)
Jul-14	1.36	Remington Plaza	59,256	1987	6,896,000	116			Strip/unanchored property; tenants: Richard Walker's Pancake House, Subway, Plato's Closet; part of 2-property portfolio
Oct-14	1.41	GolfView Center	53,349	1974	4,000,000	75			Strip/unanchored property; tenants: Rosati's Pizza, Marhaba Halal Foods; prior sale: Dec. 2005 ($4.5 mil)

Columbus Festival Plaza

Date	Distance from Columbus (miles)	Property Name	Sq. Ft.	Year Built	Price ($)	$ per Sq. Ft.	Cap Rate	Occupancy	Comments/Notes
Aug-14	2.08	Fox Run Square	143,512	1986	25,650,000	179			Strip/grocery property; tenants: Ace Hardware, GNC, UPS, Verizon Wireless, Radio Shack; prior sale: Jan. 2008 ($23.2 mil)
Oct-13	3.36	363 N Weber Rd	90,333		6,400,000	71			Strip/unanchored property; tenants: Hobby Lobby, Victory Cathedral Worship Center; prior sale: Jan. 2004 ($8.1 mil); bought for occupancy
Jan-15	3.88	The Landings	112,861	2001	9,500,000	84		49%	Strip/anchored property; tenants: OfficeMax, Smashburger, Noodles & Co; prior sale: Jan. 2003 ($22.0 mil)
Aug-14	4.56	Prairie Point SC	91,535	1998	18,151,363	198			Strip/grocery property; tenants: Dominick's, UPS Store, Hair Cuttery, Chase, Quiznos; prior sale: Jan. 2010 ($11.3 mil); part of 9-property portfolio

Table 3.14 Local sales analysis by building size (based on Chicago market retail building sales, April 2014 – March 2015)

Building Size	# of Sales	Sales Volume ($)	$ per Sq. Ft.	Cap Rate (%)
< 25,000 sq. ft.	728	1,313,849,244	223.46	7.60
25,000–99,000 sq. ft.	113	926,493,100	186.18	8.27
100,000–249,000 sq. ft.	18	290,722,709	113.66	7.79
> 250,000 sq. ft.	5	396,400,500	180.96	6.44

Source: CoStar

Table 3.15 Modeling assumptions

a. For Brookline:
   i. Hold the property another 5 years.
   ii. Operating expenses beginning in 2016 will grow at a 2% annual rate, independent of occupancy.
   iii. The property will require 17% of NOI to be spent on capital expenditures, in addition to any amounts spent on tenant improvements and broker commissions.
   iv. FedEx Ship Center will renew its lease with 75% probability. The lease will run 10 years, require tenant improvements of $12/ft^2, have a first-year rent of $26.95, and contain a 2% annual escalation. Should FedEx not renew, the space will remain vacant for 1 year before being filled by a tenant who will also lease for 10 years, have a first-year rent of $27.49, and have a 2% escalation. The new tenant would require tenant improvements of $35/ft^2. Cirano will have to pay a leasing broker $38,000 upon the signing of a new tenant to a 10-year lease. (If FedEx renews its lease, he will not need a broker.) Regardless of tenant, tenant improvements are paid in 2020.
   v. Tenants under leases will not default during the term of the lease.
b. For Columbus Festival:
   i. Hold the property another 3 years.
   ii. Operating expenses beginning in 2016 will increase by 11% due to the addition of a CVS Pharmacy, but will subsequently grow at a 2% annual rate, independent of occupancy.
   iii. The property will require 23% of NOI to be spent on capital expenditures, in addition to any amounts spent on tenant improvements and broker commissions.
   iv. Johnny B's Dry Cleaners will vacate the property at the end of 2016. That space will remain vacant in 2017, but a new tenant will lease that space for 10 years at $24.50/ft^2, with a 1% annual escalation. The new tenant will require tenant improvements of $30/ft^2. Cirano will have to pay a leasing broker $26,000 upon the signing of this new tenant. Tenant improvements and commissions are paid in 2017, but rent collected under the new lease begins in 2018.
   v. JC Penney will renew with 90% probability. If it does renew, it will sign another long-term lease beginning at $21.00/ft^2, with $0.50 increments every 5 years. If JC Penney does not renew, the best option would be to lease the space out to a series of seasonal tenants, which would bring in approximately $350,000/year in rental income. The seasonal tenants would reduce the operating expenses of the property by 4% relative to where they would be with JC Penney. There would be no tenant improvement expense with seasonal tenants, but a leasing broker charges $15,000/year to help in acquiring such tenants. No other true anchor for the space would be found during the next several years.
   vi. Currently vacant space will remain vacant.
   vii. Tenants under leases will not default during the term of the lease.

Table 3.16 Future rent rolls for Brookline Road Shopping Center

Tenant	Sq. Ft.	2016	2017	2018	2019	2020	2021
Jewel-Osco Grocery Store	35,000						
Rate		$20.50	$20.50	$20.50	$20.50	$23.50	$23.50
Income		$717,500.00	$717,500.00	$717,500.00	$717,500.00	$822,500.00	$822,500.00
Wells Fargo Bank	9,500						
Rate		$27.57	$28.12	$28.68	$29.26	$29.84	$30.44
Income		$261,900.33	$267,138.34	$272,481.11	$277,930.73	$283,489.34	$289,159.13
Chipotle	6,500						
Rate		$34.48	$35.51	$36.58	$37.68	$38.81	$39.97
Income		$224,115.13	$230,838.58	$237,763.74	$244,896.65	$252,243.55	$259,810.85
General Dentistry Office	4,500						
Rate		$29.00	$29.00	$29.00	$29.00	$29.00	$29.00
Income		$130,500.00	$130,500.00	$130,500.00	$130,500.00	$130,500.00	$130,500.00
FedEx Ship Center	10,111						
Rate		$25.39	$25.90	$26.42	$26.95	$27.49	$28.04
Income		$256,757.30	$261,892.45	$267,130.30	$272,472.90	$277,922.36	$283,480.81
Total	65,611	$1,590,772.76	$1,607,869.37	$1,625,375.14	$1,643,300.28	$1,766,655.25	$1,785,450.79

74 Debt financing

Table 3.17 Future rent rolls for Columbus Festival Plaza

Tenant	Sq. Ft.	2016	2017	2018	2019
JC Penney	35,000				
Rate		$20.50	$20.50	$21.00	$21.00
Income		$717,500.00	$717,500.00	$735,000.00	$735,000.00
Valdez Value Groceries	22,456				
Rate		$18.00	$18.00	$18.00	$18.00
Income		$404,208.00	$404,208.00	$404,208.00	$404,208.00
Subway	4,500				
Rate		$27.00	$27.00	$30.00	$30.00
Income		$121,500.00	$121,500.00	$135,000.00	$135,000.00
SportRite Sporting Goods	14,555				
Rate		$20.00	$20.00	$20.00	$20.00
Income		$291,100.00	$291,100.00	$291,100.00	$291,100.00
Johnny B's Dry Cleaners/ NEW TENANT	4,500				
Rate		$26.50	$26.50	$24.50	$24.75
Income		$119,250.00	$119,250.00	$110,250.00	$111,352.50
Italian Stallion Training Gym	8,950				
Rate		$18.80	$19.00	$19.20	$19.40
Income		$168,260.00	$170,050.00	$171,840.00	$173,630.00
Vicenzo's Pizza	4,500				
Rate		$18.57	$18.94	$19.32	$19.71
Income		$83,570.13	$85,241.53	$86,946.36	$88,685.29
First Bank Credit Union	10,111				
Rate		$24.41	$24.90	$25.39	$25.90
Income		$246,787.10	$251,722.85	$256,757.30	$261,892.45
CVS Pharmacy	13,500				
Rate		$28.00	$28.28	$28.56	$28.85
Income		$378,000.00	$381,780.00	$385,597.80	$389,453.78
Vacant units	9,000				
Rate		$0.00	$0.00	$0.00	$0.00
Income		$0.00	$0.00	$0.00	$0.00
Total gross revenue	113,572	$2,530,175.23	$2,542,352.38	$2,576,699.47	$2,590,322.02

each of his 2 properties (Table 3.18 and 3.19). Now that Cirano had become more than a skilled financial analyst, it seemed like he could readily have the numbers justify anything, so he took a moment to reflect upon his assumptions. After doing so, he had to consider whether or not a lender's willingness to lend should be the determining factor in deciding how much to borrow. Perhaps he shouldn't always leverage his investments to the maximum possible amount.

Table 3.18 Brookline pro forma cash flow

	2016	2017	2018	2019	2020	2021
Gross rental revenue	$1,590,772.76	$1,607,869.37	$1,625,375.14	$1,643,300.28	$1,766,655.25	$1,785,450.79
Vacancy					$68,118.23	
Net rental revenue	$1,590,772.76	$1,607,869.37	$1,625,375.14	$1,643,300.28	$1,698,537.03	$1,785,450.79
Total operating expenses	$681,105.70	$694,727.82	$708,622.37	$722,794.82	$737,250.72	$751,995.73
Net operating income	$909,667.06	$913,141.55	$916,752.77	$920,505.46	$961,286.31	$1,033,455.06
Capital costs						
Tenant improvements					$179,470.25	
Leasing commissions					$9,500.00	
Capital reserves	$154,643.40	$155,234.06	$155,847.97	$156,485.93	$163,418.67	$175,687.36
Total capital costs	$154,643.40	$155,234.06	$155,847.97	$156,485.93	$352,388.92	$175,687.36
Operating cash flow	$755,023.66	$757,907.49	$760,904.80	$764,019.53	$608,897.39	$857,767.70
Property sale						
Total property cash flow	$755,023.66	$757,907.49	$760,904.80	$764,019.53		
Current property valuation						

76 Debt financing

Table 3.19 Columbus pro forma cash flow

With JC Penney

	2016	2017	2018	2019
Gross rental revenue	$2,530,175.23	$2,542,352.38	$2,576,699.47	$2,590,322.02
Vacancy		$119,250.00		
Net rental revenue	$2,530,175.23	$2,423,102.38	$2,576,699.47	$2,590,322.02
Total operating expenses	$1,230,253.40	$1,254,858.47	$1,279,955.64	$1,305,554.75
Net operating income	$1,299,921.83	$1,168,243.91	$1,296,743.83	$1,284,767.27
Capital costs				
Tenant improvements		$135,000.00		
Leasing commissions		$26,000.00		
Capital reserves	$298,982.02	$268,696.10	$298,251.08	
Total capital costs	$298,982.02	$429,696.10	$298,251.08	
Operating cash flow	$1,000,939.81	$738,547.81	$998,492.75	
Property sale				
Total property cash flow	$1,000,939.81	$738,547.81		
Current property valuation				

Without JC Penney

	2016	2017	2018	2019
Gross rental revenue	$2,530,175.23	$2,542,352.38	$2,191,699.47	$2,205,322.02
Vacancy		$119,250.00		
Net rental revenue	$2,530,175.23	$2,423,102.38	$2,191,699.47	$2,205,322.02
Total operating expenses	$1,230,253.40	$1,254,858.47	$1,228,757.41	$1,253,332.56
Net operating income	$1,299,921.83	$1,168,243.91	$962,942.05	$951,989.46
Capital costs				
Tenant improvements		$135,000.00		
Leasing commissions		$26,000.00	$15,000.00	
Capital reserves	$298,982.02	$268,696.10	$221,476.67	
Total capital costs	$298,982.02	$429,696.10	$236,476.67	
Operating cash flow	$1,000,939.81	$738,547.81	$726,465.38	
Property sale				
Total property cash flow	$1,000,939.81	$738,547.81		
Current property valuation				

Notes

1. The amortization period is the length of time required to reduce the outstanding loan balance to 0.
2. Numbers in the text may reflect rounding. See the accompanying spreadsheet for exact calculations.
3. Non-recourse loans typically become recourse if the default is shown to be the result of investor misrepresentation or malfeasance such as diverting rents or failing to pay real estate taxes.
4. If the mortgage contains a due on sale clause, the full outstanding balance of the loan comes due immediately.
5. This is not strictly true as it assumes that the relevant interest rate is 3% for all of the 36 remaining months. If the Treasury yield curve is not flat at 3%, then this sum may be more or less than required to replicate the original cash flow. Unlike yield maintenance, defeasance penalties explicitly incorporate the slope of the yield curve.
6. Typically the borrower will be a special purpose entity established to own the investment property. So technically, the lender is reviewing the creditworthiness of the sponsor of the entity that owns the collateral property.

7 A lockout provision is a clause in a mortgage contract that prevents the borrower from prepaying the loan during a specified time period.
8 The DSCR is the ratio of property NOI divided by required debt service (both principal and interest).
9 This means a 5% prepayment penalty after 5 years of payments, a 4% penalty after 6 years of payments, and so on, decreasing 1 percentage point each year.
10 The project required $1 million in January 2014, another $500,000 in July 2014, and the remaining $250,000 in July 2015. Interest on the construction loan was accruing monthly, which led to an outstanding loan balance of $1,922,108.08 to be repaid on December 31, 2015.

4 Equity partnerships

Introduction

In Chapter 3, we saw that an investor may wish to borrow money to finance, in part, the acquisition of an investment property. We also saw that the underwriting done by a lender may influence the amount of money that can be borrowed. In the example we've been following, an investor identified the opportunity to purchase an industrial property that, according to our analysis, has a discounted cash flow (DCF) valuation of $4,756,166. To help with the financing of this property, our assumption in Chapter 3 was that the investor borrowed $3,000,000. The purpose of Chapter 4 is to consider situations where the investor does not have sufficient equity to fund the remainder of the purchase price. Specifically, consider the case where our investor has negotiated a purchase price for the 2-space industrial building of $4,600,000. Because the purchase price is less than the DCF valuation, this would be a positive net present value (NPV) transaction and one that we would encourage our investor to pursue. However, after borrowing $3,000,000 from a bank, our investor does not have the $1,600,000 still required for the purchase. Equity partnerships may be the solution to this problem for our investor.

The partners

This chapter considers situations where a property investor has the ability to locate and manage property investment opportunities. However, such investors may have insufficient wealth to acquire the investment opportunities by themselves. In such circumstances, property investors may be well-suited to be the general partner (GP) of a **real estate private equity partnership**. **General partners** are also sometimes referred to as deal sponsors. Essentially, the GP/sponsor seeks to "sell" its proficiency in real estate investing to others. Which others? Well, certainly not others like them. That is, GPs wouldn't necessarily want to partner with others who have the expertise in identifying investment opportunities and managing property, but who are limited in wealth. Rather, they would ideally like to find people that would love to invest in property and have enough wealth to do so, but who lack either the interest or ability to identify opportunities and/or to manage properties themselves. Investors such as this would make excellent limited partners (LPs). **Limited partners** (also sometimes simply referred to as outside investors) contribute their wealth into the GP's investment idea, but are neither involved with identifying opportunities nor involved with the day-to-day management of the property. When a GP and one or more LPs pool their equity to acquire an investment property, it is called a real estate private equity partnership. This partnership typically would form a joint venture (JV) for the purpose of acquiring an investment property. This structure is illustrated in Figure 4.1.

Equity partnerships 79

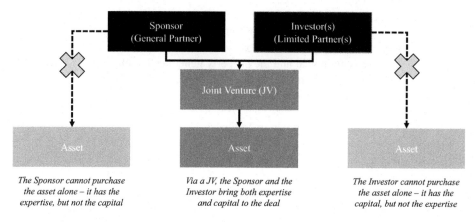

Figure 4.1 Structure of a private equity partnership

JV structure basics

Would a GP and 1 or more LPs be willing to become partners if the investments of both groups of partners were treated on a **pari passu** basis? Pari passu means that cash flows from an underlying investment are distributed to members of a partnership in proportion to the size of their equity investments. Note that each partnership group is necessary for the investment. The GP is bringing the deal and the expertise. The LPs are bringing (most of) the money. So conceptually, it is possible that partnership cash flows could be treated pari passu. In practice, however, GPs typically provide only a fraction (typically between 5% and 15%) of the equity needed for any given investment. As a result, LPs are concerned with the GP's incentives. That is to say, because the GP has so little invested in any given property, LPs may rightfully worry that the GP will not exert sufficient effort to make the deal successful. As a result, LPs wish to give their GP a financial incentive to ensure that they do. Therefore, real estate private equity partnership agreements generally call for the GP to earn a **promote**. A promote refers to a situation where the GP receives cash flows in excess of its pari passu share. To provide the right incentives, LPs generally insist that the GP only earns a promote after satisfying some benchmark return, or **hurdle rate**. The hurdle rate is the rate of return earned by the equity in the deal *beyond which* the GP earns a promote. Thus, real estate private equity partnership agreements typically call for pari passu cash flow distributions until a hurdle rate is surpassed, at which point the GP begins to earn a promote.

A one-period example

Before turning to more realistic examples in the next section, it is perhaps useful to illustrate hurdles and promotes with a stylized, one-period example.

> Example 4.1: Suppose that a GP identifies an excellent property investment that can be acquired for $5,000,000. A bank underwrites the property and provides a $3,250,000 mortgage that requires annual, interest-only payments at a rate of 5%. That leaves $1,750,000 to be provided by equity investors. Suppose the GP provides $175,000, or 10% of the equity commitment. The GP then finds 1 or more LPs willing to provide

80 Equity partnerships

$1,575,000. If the GP succeeds in finding the LP money, then the deal can proceed. Suppose the terms of the deal that were necessary to get both the GP and LP to agree to proceed called for pari passu distributions up to an 8% hurdle rate, beyond which the GP would earn a 20% promote. Suppose that with these terms in place, the GP manages the property for a year and the property generates cash flow of $375,000. The GP then sells the property at the end of the year for $5,100,000. Calculate the return on the property, debt, equity, GP equity, and LP equity.

The solution to this example is shown in Table 4.1.

Note that the property generated $375,000 in cash flow and $100,000 in capital gain. In total, therefore, the property generated a 9.5% return ($475,000 ÷ $5,000,000). This can be decomposed into a 7.5% cash flow yield and a 2% return on price appreciation. The return on the property's debt is easily calculated to be 5%. The mortgage was interest only, which means that the loan required a repayment of $3,412,500. By construction, this generates a 5% return for the lender ($3,412,500 − $3,250,000) ÷ $3,250,000 = 5%.

Turning next to the equity component of this deal, the property required $1,750,000 in equity upon acquisition. At the end of the deal, equity receives $2,062,500, which is the difference between the total cash flow received ($5,100,000 + $375,000) and what is repaid to the lender ($3,412,500). Therefore, total return to equity in this deal is 17.9% = ($2,062,500 − $1,750,000) ÷ $1,750,000.

In a private equity limited partnership, the GP and LP will typically not receive the same return on their investments unless all cash flows are distributed pari passu. In this instance,

Table 4.1 1-period example of equity splits

Property price	$5,000,000	
Bank loan	$3,250,000	
Equity required	$1,750,000	
GP equity	$175,000	10%
LP equity	$1,575,000	90%
Hurdle rate	8%	
Promote	20%	
		Rate of return
Cash flow	$375,000	7.5%
Property sale	$5,100,000	2.0%
Property return		9.5%
Debt repayment	$3,412,500	5%
Equity cash flow	$2,062,500	17.9%
GP with hurdle	$189,000	
LP with hurdle	$1,701,000	
Excess cash	$172,500	
GP promote	$51,750	
LP share	$120,750	
Total equity cash flow		Rate of return
GP	$240,750	37.6%
LP	$1,821,750	15.7%

the $2,062,500 returned to equity will not be distributed pari passu. The partnership agreement specifies how this remaining cash flow is to be split between the GP and LPs. The hurdle rate was set at 8%, so cash flow up to $1,890,000 = (1 + 8%) × $1,750,000 is shared pari passu. This means that the first $1,890,000 of equity cash flow is split in the same 10%/90% proportion that the initial equity investment was made. So $189,000 is given to the GP and $1,701,000 is given to the LPs. After this money has been paid, there still remains $172,500 of cash flow to equity in excess of the hurdle rate. Since the hurdle has been surpassed, the GP earns a 20% promote, which means that the GP will take an *additional* 20% of the cash flow in excess of the 10% it contributed. Thus, the GP will receive an additional $51,750 = 30% × $172,500.

In total, therefore, at the end of the investment, the GP received a cash flow of $240,750 = $189,000 + $51,750 and the LPs received a cash flow of $1,821,750 = $1,701,000 + $120,750. These cash flows translate into a rate of return of 37.6% for the GP and 15.7% for the LPs. Thus, the GP's return was higher than the return on equity and the LP's return was lower. This is typical in a partnership where returns exceed the hurdle rate.

These calculations illustrate how cash flows in this one-period example would be split between the GP and the LPs if the underlying equity delivered a 17.9% return. Of course, many return outcomes are possible. Figure 4.2 graphs the return to the GP and the return to the LPs as a function of the overall return on equity. Note that the return to both partners is identical whenever the return on overall equity is less than or equal to the hurdle rate of 8%. This makes sense, because for rates of return less than the hurdle, the GP does not earn a promote. Above the hurdle rate, a wedge opens up between the GP and LP returns, with the GP generating progressively higher returns. This additional return for the GP was the incentive given to the GP to ensure the investment was a success.

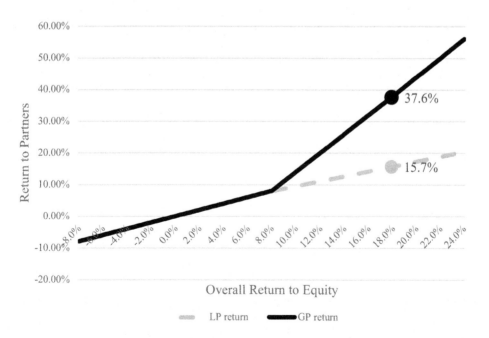

Figure 4.2 Partnership returns as a function of equity returns

82 Equity partnerships

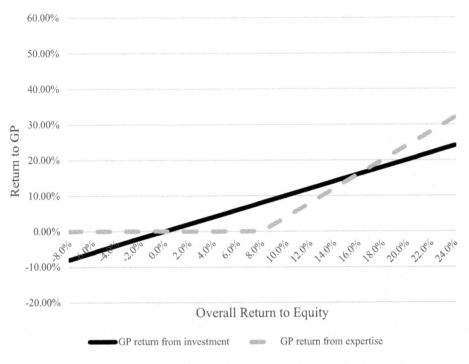

Figure 4.3 Decomposition of GP returns

Another way to view the returns to a GP in this partnership is to deconstruct the GP return into 2 parts. The first component of GP returns is the return received simply by investing money into the deal. As illustrated in Figure 4.3, the solid line has a slope of 1, indicating that the GP earns additional return when overall equity earns a higher return. The dashed bent line represents the return to the GP earned in excess of the overall equity return. As illustrated, the excess return is 0 until the project equity return exceeds the hurdle rate. Beyond the hurdle rate, the slope of the dashed line is 2 because the GP receives 30% of the excess cash over the hurdle return when it had only invested 10% of the equity. The dashed kinked line has the shape of a call option. Thus, intuitively, one can view the investment of the GP as the combination of an equity investment and a long call option with a strike price equal to the hurdle rate.

A multiple-period example

In reality, when a partnership acquires a property, the investment lasts several years. In this section, we extend the intuition provided in the previous section to learn more about the mechanics of real estate private equity partnerships in a more realistic investment environment. To do so, we return to the example of the 2-tenant industrial building that we have used throughout the book. Recall that in Chapter 3, we showed how to split cash flow of the property into debt cash flows and equity cash flows. We did this by subtracting all payments to the lender from total property cash flow. Table 4.2 replicates Table 3.3, which shows the last few lines of our pro forma model.

Table 4.2 Multiple period cash flows to property, debt, and equity

	Year 0	Year 1	Year 2	Year 3	Year 4	Year 5
Net operating income (NOI)		$184,680	$405,240	$337,280	$481,460	$477,822
Building improvements		$0	$0	$80,000	$0	$0
TIs		$60,000	$0	$60,000	$0	$0
Leasing commissions		$165,750	$0	$180,000	$0	$0
Holding period cash flow		-$41,070	$405,240	$17,280	$481,460	$477,822
Reversion cash flow						
Sales price						$5,926,615
Sales commission						$177,798.46
Total property cash flow	-$4,600,000	-$41,070	$405,240	$17,280	$481,460	$6,226,639
Total debt service	-$3,000,000	$195,026	$195,026	$195,026	$195,026	$2,819,580
Total cash flow to equity	-$1,600,000	-$236,096	$210,214	-$177,746	$286,434	$3,407,059

Property return	9.52%
Debt return	4.20%
Equity return	16.26%

Suppose that this industrial property was not purchased by a sole investor. It may be the case that our investor does not have $1,600,000 to invest in the property and therefore raises 90% of the equity required from limited partners. These partners establish a JV to acquire the property and hold it for 5 years. As illustrated in Table 4.2, the ownership of this industrial property requires equity investments upon purchase and again in Years 1 and 3. Equity investors will receive cash flow in Years 2, 4, and 5. Which partners are responsible for which investments and which will receive which cash flows?

The private equity **waterfall**, detailed in the partnership agreement, will be our guide to understanding how the equity level cash flows will be shared between partners. A waterfall is the process by which one determines how any given cash flow in a real estate private equity investment is split between the GP and the LPs. Because partnership agreements are negotiated privately between partners, details of these agreements may vary considerably across investments. In this section, we will consider 2 examples of how waterfalls may be structured, but keep in mind *any* mechanism to split cash flows between partners is possible. In reality, there are many variations on the deal structures that we describe in this chapter.

Return of capital waterfall

To begin, let's assume that the JV that acquired the industrial building has a partnership agreement that provides the waterfall details outlined in Table 4.3. This waterfall must specify sufficient details such that *any* outcome of the underlying investment can be addressed. We will then split the equity level cash flows we calculated in Chapter 3 between the GP and LPs according to the waterfall details.

In this example, the pari passu split is 90% by the LPs and 10% by the GP. This was the agreed upon equity financing of the property. This means that the LPs are responsible for contributing $1,440,000 and the GP is responsible for contributing $160,000 towards the acquisition of the property.

At this stage, it is necessary to introduce some more bookkeeping, which is necessary to keep track of the LP investment separately from the GP investment. Table 4.4 presents the

84 Equity partnerships

Table 4.3 Return of capital waterfall

Deal terms	
Contribution by LPs	90%
Contribution by GP	10%
Preferred return	8%
Promote	20%
Subsequent investments: pari passu	
Distributions	
Pari passu	Until all capital is returned
70/30	All excess cash flow

Table 4.4 Partnership equity accounts, return of capital

	Year 0	Year 1	Year 2
LP equity account			
Beginning of year balance	$0	$1,440,000	$1,767,686
Additional investment	$1,440,000	$212,486	$0
Preferred return earned		$115,200	$141,415
Return of capital		$0	$189,193
LP share of excess cash flow		$0	$0
End of year balance	$1,440,000	$1,767,686	$1,719,908
GP equity account			
Beginning of year balance	$0	$160,000	$196,410
Additional investment	$160,000	$23,610	$0
Preferred return earned		$12,800	$15,713
Return of capital		$0	$21,021
GP share of excess cash flow		$0	$0
End of year balance	$160,000	$196,410	$191,101

first 2 years of the **equity accounts** of the investors. Following investment, the LPs have invested $1,440,000 and the GP has invested $160,000. In Year 1 of the investment, both the LP and GP have earned a **preferred return**. A preferred return – often shortened to "the pref" by experienced private equity real estate investors – is a return given to investors on equity invested into the deal that takes *preference* to the receipt of any cash flow. That is, investors must receive their preferred return before any of their investment is returned. The preferred return can be thought of as a minimum "interest rate" that investors in the deal earn. However, this rate is not guaranteed and is only received if the underlying property cash flow is sufficient. In the waterfall outlined in Table 4.3, the preferred return is 8%. The line in the equity account *preferred return earned* calculates 8% of the beginning of the year equity balance. Therefore, the LPs have earned a pref of 8% × $1,440,000 = $115,200. Similarly, the GP has earned a pref of $12,800 = 8% × $160,000.

Having *earned* a preferred return in Year 1 does not mean that the partners actually *receive* a preferred return in Year 1. In this example, they won't. That is because in Year 1 of the investment, the cash flow to equity is negative. Recall from Chapter 3 that this occurs because of the significant vacancy and concessions granted in the first year of operation. Negative equity cash flow means that equity investors – in this case, the partnership – must put additional cash into the property. A **capital call** is when investors are asked for additional funds during the life of an investment.[1] According to the waterfall described in Table 4.3, subsequent investments (negative equity level cash flows) are treated on a pari passu basis. Therefore, LPs make an additional contribution of $212,486 while the GP invests an additional $23,610. Because there is no positive cash flow to distribute to the partners in the first year, we can calculate each investor's equity balance at the end of the year as the sum of (1) their initial investment, (2) the preferred return that was earned but not paid, and (3) the additional investment necessary in the first year.

In Year 2 of the investment, we can again calculate the preferred return earned by the investors. The preferred return is higher in Year 2 because the equity investment has grown. In the second year, however, there is positive equity cash flow that can be distributed to investors. According to the waterfall, positive cash flow is distributed pari passu until all investor capital has been returned. This means that the LPs receive $189,193 = 90% × $210,214 and the GP receives $21,021 = 10% × $210,214. This waterfall agreement specifies that cash returned to investors is considered a return of capital. Therefore, the investors' equity balance at the end of the second year is calculated as the initial equity balance + preferred return earned − return of capital received.

The waterfall also specifies the conditions under which the GP will earn a promote. The 20% promote is earned only after all capital has been returned. It is typically the case that capital in a real estate private equity partnership is not fully returned until reversion, when the underlying property is sold. That is also the case in this example. The split of equity cash flow in Year 5 is detailed in Table 4.5. After the property is sold and the outstanding balance on the mortgage loan is repaid, there is a cash flow to equity of $3,407,059 in Year 5. Of this amount, $2,074,765 is required to pay the preferred return and return all of the capital of the LPs, and $230,529 is required to do the same for the GP. This leaves $1,101,764 of excess cash to be distributed to the partnership. Now, according to the waterfall, cash flow received

Table 4.5 Cash flows upon reversion, return of capital

	Year 5
LP equity account	
Beginning of year balance	$1,921,079
Additional investment	$0
Preferred return earned	$153,686
Return of capital	$2,074,765
LP share of excess cash flow	$771,235
End of year balance	$0
GP equity account	
Beginning of year balance	$213,453
Additional investment	$0
Preferred return earned	$17,076
Return of capital	$230,529
GP share of excess cash flow	$330,529
End of year balance	$0
Total LP cash flow	$2,846,000
Total GP cash flow	$561,059
LP IRR	14.62%
GP IRR	27.98%

86 *Equity partnerships*

after all capital has been returned to investors is split 70%/30%, with the GP earning the 20% promote on its 10% pari passu share. This final split delivers an additional $771,235 = 70% × $1,101,764 to the LPs and an additional $330,529 = 30% × $1,101,764 to the GP.

Having calculated how the equity cash flow is split between partners in each year of the pro forma, one can calculate the returns to the partners and compare those returns to the return to equity overall. Recall that in this example the return to equity investors was 16.26%. However, when equity is funded by this particular private equity real estate partnership, the GP earns 27.98%, while the LPs earn 14.62%. As was the case in our one-period example, the GP earns higher returns than the LP because the underlying equity investment returned higher than the level of the hurdle rate, which in this example was the preferred return of 8%.

Return on capital waterfall

In the previous example, the waterfall agreement designated that all cash flows delivered to the partners be treated as a return of capital, thereby reducing the level of equity investment in the deal. In other partnership agreements, cash flow to partners is treated as a return *on* capital, with cash flows in excess of what is necessary to pay the preferred return being eligible for a promote. Table 4.6 describes how a return on capital waterfall might look.

Let's work through the same industrial building example with this new waterfall. The cash flow splits in the first 2 years are shown in Table 4.7. In Year 1, the waterfall is identical to what we saw before. Investors receive a capital call to fund the needed cash infusion, and the capital call is allocated on a pari passu basis. At the end of Year 1, the LPs earn a preferred return of $115,200.00, which is the same preferred return earned in the previous waterfall. Similarly, the GP also continues to earn a preferred return of $12,800.00. However, unlike the previous waterfall where LPs received 90% of cash flow in Year 2, investors are receiving a return *on* their capital and not a return *of* their capital before excess cash flow earns the GP a promote. That means that of the $210,214.28 that equity receives in Year 2, the LP receives only $178,575.53, which can be calculated as the sum of its preferred return ($141,414.89) and 70% of the excess cash after all preferred returns have been paid ($37,160.64 = 70% × [$210,214.28 − $141,414.89 − $15,712.77]).[2] In Year 3, there is another capital call and the

Table 4.6 Return on capital waterfall

Deal terms	
Contribution by LPs	90%
Contribution by GP	10%
Preferred return	8%
Promote	20%
Subsequent investments: pari passu	
Distributions during holding period	
Pari passu	Until preferred return is paid
70/30	All cash flow in excess of preferred return
Distributions at reversion	
Pari passu	Until preferred return is paid and all capital is returned
70/30	All excess cash flow

Equity partnerships 87

Table 4.7 Partnership equity accounts, return on capital

	Year 0	Year 1	Year 2
LP equity account			
Beginning of year balance	$0.00	$1,440,000.00	$1,767,686.14
Additional investment	$1,440,000.00	$212,486.14	$0.00
Preferred return earned		$115,200.00	$141,414.89
Preferred return paid		$0.00	$141,414.89
Return of capital			
LP share of excess cash flow		$0.00	$37,160.64
End of year balance	$1,440,000.00	$1,767,686.14	$1,767,686.14
GP equity account			
Beginning of year balance	$0.00	$160,000.00	$196,409.57
Additional investment	$160,000.00	$23,609.57	$0.00
Preferred return earned		$12,800.00	$15,712.77
Preferred return paid		$0.00	$15,712.77
Return of capital			
GP share of excess cash flow		$0.00	$15,925.99
End of year balance	$160,000.00	$196,409.57	$196,409.57

Table 4.8 Cash flows upon reversion, return on capital

	Year 5
LP equity account	
Beginning of year balance	$2,069,072
Additional investment	$0
Preferred return earned	$165,526
Preferred return paid	$165,526
Return of capital	$2,069,072
LP share of excess cash flow	$646,921
GP equity account	
Beginning of year balance	$229,897
Additional investment	$0
Preferred return earned	$18,392
Preferred return paid	$18,392
Return of capital	$229,897
GP share of excess cash flow	$277,252
Total LP cash flow	$2,881,519
Total GP cash flow	$525,540
LP IRR	14.59%
GP IRR	28.00%

necessary capital contributions are shared pari passu between the partners as was done in Year 1. Year 4 cash flow splits are then calculated as was done in Year 2.

In Year 5, the waterfall calls for investors to receive their capital back before excess cash flow can be promoted. Therefore, the LP cash flow in Year 5 can be calculated as $2,881,518.68, which is the sum of the preferred return paid ($165,525.77 = 8% × $2,069,072.18), the return of the LP capital ($2,069,072.18), and 70% of the excess cash flow remaining ($646,920.72 = 70% × [$3,407,059.08 − $165,525.77 − $18,391.75 − $2,069,072.18 − $229,896.91]). These additional cash flow splits are shown in Table 4.8.

Note that with this waterfall agreement, the GP does slightly better relative to the return of capital example. This is because the GP is able to earn a (small) promote *prior* to all investors receiving their capital back. So all else equal, a return *of* capital deal is typically better for the LPs than a return *on* capital deal. However, in reality, all else need not be equal. That is, a return on capital waterfall might give enhanced returns to LPs if they successfully negotiate other terms of the deal more favorably, for instance, negotiating a lower promote for the GP.

Fee structures in private equity real estate partnerships

The promote is one way in which LPs pay the GP for finding the deal, arranging any debt financing, raising money from LPs, and managing the property during the holding period. The partnership agreement typically also includes a variety of other fees, which we have not included in our example waterfalls. Limited partners often pay an **acquisition fee** in recognition of the GP's work to find the transaction. **Asset management fees** are paid by LPs to compensate a GP for managing the joint venture, which includes handling the financial reporting for the LPs and other duties associated with the effective management of the JV. These fees are typically charged as a percentage of equity invested or the asset value of the property. A **property management fee** covers the costs associated with having an onsite property manager that oversees the investment on a day-to-day basis. Unlike the previous 2 fees, the property management fee is paid by the entire JV because management expenses would appear on the property pro forma and thus would be deducted before any cash flow is received by the equity partners. Property management fees would typically be charged as a percentage of gross rental revenue or some other metric of property-related income.

Real estate private equity funds

This chapter presented a real estate private equity joint venture as a way that a property investor could raise money from equity partners to finance the acquisition of a single property. Suppose that the investor had access to many deals? One solution would be to establish a number of JVs, each of which invested in a single asset. In this instance, a sponsor would have a number of deals in which it serves as the GP. Potential LPs benefit from sponsors asking for money on a deal-by-deal basis because that allows the LPs to evaluate the merits of each investment before deciding whether or not to invest. LPs can choose to invest in some properties but not others.

From the sponsor's perspective, raising money on a deal-by-deal basis is problematic for the same reasons that it is beneficial to the LPs. It takes time and effort to raise money, and to have to raise money for each property investment can become excessively costly. Therefore, a sponsor with potentially many investment opportunities might instead prefer to raise money once, by establishing a **real estate private equity fund**. Unlike in a JV structure where the sponsor has identified to potential LPs the specific property being acquired, a prospectus to a fund will instead highlight (1) who the sponsor (GP) is, (2) what track record the sponsor has in previous investments, (3) who and/or what type of limited partners the GP has worked with in the past, (4) what the investment criteria are for identifying specific fund investments, and (5) what investors should expect in terms of returns on their investment. A fund structure therefore allows a GP to raise money once, yet invest in many properties. Of course, the ability of a sponsor to raise a fund depends crucially on whether or not LPs are willing to commit capital *before* seeing specific properties for investment. Ultimately, the ability to raise a fund has to do with the sponsor's reputation and track record, which is why a fund prospectus typically highlights the sponsor's past successes. Successful sponsors of real estate private equity JVs may be able to successfully raise a fund. New or otherwise unproven sponsors likely will have to raise money on a deal-by-deal basis using JVs, at least until they establish a record of proven performance.

A real estate private equity fund's typical structure is shown in Figure 4.4. The sponsor raises a fund from limited partners and then uses the commitments from partners to invest in multiple properties. Typically, the fund prospectus will put restrictions on the actions of the GP. Restrictions often include (1) limits to how long the sponsor has to invest committed capital, (2) what happens to uninvested capital, (3) the target holding period for assets acquired, and (4) the expected life of the fund. A stylized timing of such a fund is given in Figure 4.5.

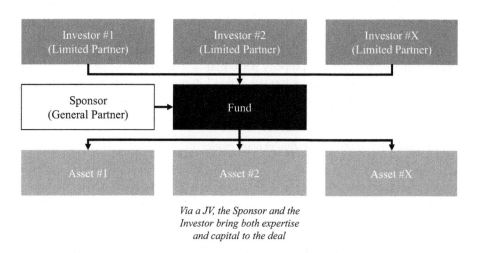

Figure 4.4 Real estate private equity fund structure

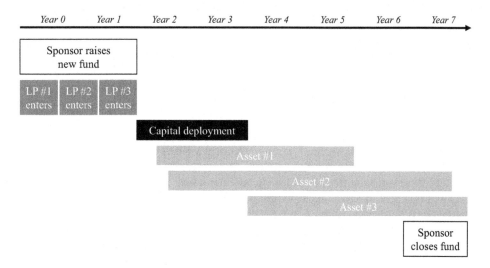

Figure 4.5 Timing of a real estate private equity fund

As was the case in joint ventures, investors in real estate private equity funds will similarly be assessed acquisition fees, property management fees, and fund management fees. In the case of fund investing, though, management fees are typically calculated as a percentage of committed capital during the fundraising period and then switch to be based on equity deployed or gross or net asset value of the fund once the capital is put to work.

Real estate private equity: Wildcat Capital Investors

In the Wildcat case, you are put in the shoes of Jessica Zaski, the analyst for the real estate private equity firm Wildcat. Zaski is preparing the financial projections that will be used to show potential limited partners about an investment opportunity. In completing this case, you will be able to:

1 Construct a private equity waterfall under a number of different assumptions regarding partnership agreements and property performance.
2 Understand and evaluate the information that LPs are typically provided when making an investment decision for a real estate joint venture.

Wildcat Capital Investors: real estate private equity

"Okay. Now we're even," said the voice on the telephone. As he hung up the phone, James Tripp, managing director of Wildcat Capital Investors, thought back to that beautiful summer evening 2 years earlier when he was about to enter Ravinia Park to enjoy a performance of the Chicago Symphony with his friend, commercial real estate broker Katherine O'Brien. The sound of scraping metal had caught Tripp's attention just in time for him to save O'Brien's life – or so he liked to claim – by blocking an approaching bicyclist headed straight for O'Brien in a reckless attempt to cross the track ahead of an oncoming train. At the time Tripp had joked, "Now you owe me." Referring to the opportunity to purchase a piece of commercial property before the sale became public knowledge, he continued, "How about showing me a great off-market deal some day?" Now, in September 2009, it seemed that O'Brien had indeed returned the favor.

The opportunity O'Brien had just briefly outlined on the phone sounded perfect for Wildcat. Financial Commons was a 90,000-square-foot office property located in the Chicago suburb of Skokie. The building was 90% occupied and was being offered for what seemed like an incredible price of $10,400,000. Given the bleak commercial market environment at the time, such opportunities were few and far between.

But Tripp knew there were many factors that could spoil this deal. As they did with several properties each week, Tripp and Wildcat's MBA-student intern, Jessica Zaski, would have to dig deeper into the numbers. What were the economic fundamentals in the market? Who were the tenants of Financial Commons? Would Wildcat be able to profitably exit this deal in 3 to 5 years? Could it get the returns its investors demanded? And, in the midst of the worst credit crunch in Tripp's memory, would Wildcat be able to obtain financing?

Tripp called Zaski into his office. Her task would be to research the Skokie office market to derive the realistic assumptions necessary to calculate the returns Wildcat could hope to achieve by acquiring Financial Commons.

About Wildcat

Evanston, Illinois–based Wildcat Capital Investors, LLC, was a privately held entrepreneurial real estate firm that invested in a wide range of real estate assets on a deal-by-deal basis. Tripp and his friend William Paris had founded the company in 2005 to serve the growing real estate alternative investment space by raising capital from wealthy individuals and investing in commercial real estate not only through traditional equity investments but also throughout the capital structure.

Wildcat's early deals involved mezzanine loans to support multifamily development efforts. It developed a niche in providing such debt on some high-profile projects in Chicago's South Loop. At the beginning of 2009, however, Wildcat had determined that the best opportunities were in acquisitions and began exiting its existing mezzanine deals at a profit. At about the same time, the market for traditional real estate mezzanine finance dried up and many debt providers to commercial real estate lost everything as they mistakenly shared the common market assumption that rents and prices would rise forever. Thus, Wildcat had the financial resources to pursue acquisitions. Financial Commons would be its first.

The property

Through her research, Zaski learned that Financial Commons was a 3-story, Class B+ office building located in the Chicago suburb of Skokie, Illinois. Zaski was pleased to find out that although Financial Commons had been built in 1981, it had undergone a major renovation in 2007. The property sat on 6.5 acres of land, with approximately 85,812 square feet of rentable space and a large parking lot. In the fall of 2009, there were 6 tenants filling 90.5% of the rentable space. The building also had 3 small equal-size vacancies totaling 8126 square feet.

The market

Zaski knew that ensuring a good investment was about more than just the building attributes alone. She next investigated the financial health of the Chicago area and Skokie's future prospects. She also wanted to make sure Financial Commons was well placed geographically so that it was accessible and appealing to tenants and that it was not located in an oversupplied market.

Through market research and Tripp's discussions with local leasing brokers and investment sales brokers, Zaski found that the Near North submarket of Chicago, of which Skokie was a part, contained approximately 14 million square feet of corporate office space occupied by strong companies, including national businesses such as Peapod and Illinois Tool Works. Although Chicago had suffered a slight population decline over the previous decade, Skokie and its immediate surroundings had seen slow but steady population growth. Further, the submarket was one of the best-performing regions outside the downtown area of Chicago in terms of its low vacancy rates and high quoted rents (see Table 4.9).

In terms of location and appeal to high quality office tenants, Financial Commons was well positioned. The property was in an established business park that was home to dozens of employers, primarily service and professional firms. The trade area encompassed some of Chicago's most affluent zip codes and one of the region's strongest residential markets. The building was also less than half a mile from a 2.7-million-square-foot super-regional mall, Westfield's Old Orchard.

Equity partnerships

Table 4.9 Performance of office submarkets of Chicago, second quarter 2009

Market	Number of Buildings	Total Rentable Building Area (sq. ft.)	Vacancy (%)	YTD Net Absorption (sq. ft.)	Quoted Lease Rates (per sq. ft. per year)
Central Loop	104	42,229,795	13.10	−653,538	$29.31
Central North	702	33,053,460	13.80	−626,885	$20.55
Central Northwest	357	7,697,348	15.50	−78,626	$21.98
Cicero/Berwyn area	99	1,194,674	11.20	−22,149	$18.46
East Loop	84	27,729,351	16.30	−1,300,203	$27.47
Eastern East/West corridor	834	33,079,259	17.40	−428,717	$21.11
Far North	216	4,533,498	11.70	−73,701	$18.27
Far Northwest	795	9,656,432	20.80	−138,397	$19.09
Far South	183	3,200,489	13.70	4,267	$16.61
Gold Coast/Old Town	28	566,082	10.10	−7,554	$21.81
Indiana	573	7,533,197	11.10	4,401	$16.76
Joliet/Central Will	591	8,759,302	16.70	287,561	$21.71
Kenosha East	9	51,540	6.50	3,425	$25.22
Kenosha West	77	1,036,219	17.50	2,034	$14.77
Lincoln Park	171	3,232,657	8.80	−39,890	$21.52
Melrose Park area	90	2,257,568	18.30	−41,028	$14.82
Near North	414	14,283,588	9.70	−151,522	$21.80
Near South Cook	291	3,392,824	11.60	−10,970	$19.17
North DuPage	257	7,702,324	22.00	−377,563	$21.39
North Michigan Avenue	103	15,399,929	11.10	−261,687	$31.57
Northwest city	847	14,967,070	8.90	−41,017	$18.47
Oak Park area	140	1,944,237	9.50	9,037	$21.20
O'Hare	415	18,823,485	23.60	−693,419	$21.47
Porter County	267	2,283,660	13.90	−13,146	$17.49
River North	217	17,729,379	12.10	691,561	$28.13
River West	128	4,907,923	13.10	−114,128	$18.59
Schaumburg area	761	34,811,302	19.60	−1,197,740	$19.55
South Chicago	472	11,372,599	9.80	−56,127	$17.70
South Loop	36	3,452,173	5.00	40,388	$19.58
South Route 45	265	3,768,087	13.70	−44,416	$18.92
West Loop	142	49,207,787	14.40	−966,617	$30.55
Western East/West corridor	1,334	35,588,827	16.60	−205,852	$20.00
Totals	11,002	425,446,065	15.00	−6,502,218	$23.44

Source: CoStar Group.

The deal structure

Zaski knew that Wildcat, like most commercial real estate investors, would want to use a combination of debt and equity to finance the acquisition of Financial Commons. She knew that in the fall of 2009 a bank loan, especially a commercial mortgage, was unlikely due to the freeze of the credit markets. National and regional banks had all been burned by real estate and construction lending as defaults climbed. They did not have the appetite for even low-risk property lending, and as a result loan origination had recently dipped to an all-time low (see Figure 4.6). Consequently, sales of commercial property had also collapsed, which made finding comparable sales – something buyers and sellers relied on to value their properties – a difficult task (see Figure 4.7).

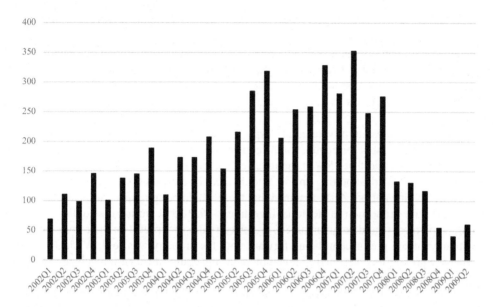

Figure 4.6 Commercial/Multifamily Mortgage Bankers Originations Index
Note: 2001 Quarterly Average = 100
Source: Mortgage Bankers Association.

Figure 4.7 Office property sales ($ billions)
Source: Real Capital Analytics, Inc.

94 Equity partnerships

She began by investigating a loan through Commercial Bridge Finance, a Chicago-based mortgage broker that specialized in this market. Based on its market knowledge and relationships with various non-bank lenders, the broker felt confident Wildcat could obtain a non-recourse 65% loan-to-value (LTV) loan from a life insurance company at a rate of 6.75%. The loan would carry a 5-year maturity and would amortize based on a 25-year schedule. The loan would require monthly payments and – as was becoming typical in the tight credit markets – would carry a 3% penalty on all prepaid balances if the loan were fully repaid before maturity.

Like most real estate private equity firms, Wildcat did not want to put much of its own cash into the deal, so it raised most of the equity for its investments from limited partners. As such, Zaski assumed that 95% of the required equity investment would come from outside investors, with Wildcat putting up 5%. Should there be any additional cash inflows required over the holding period, Zaski assumed that these would be split on a pari passu basis – that is, in the same proportion as the initial investment – 95%/5%. During the holding period, the limited partners would be given an 8% preferred return (or "pref") on their invested capital, with any unpaid pref accumulating forward until the property was sold. Wildcat would not receive a pref on its investment. Should there be any equity-level cash received in excess of the limited partners' 8% pref, it would be split between the outside investors and Wildcat at a rate more favorable to the sponsor (known as a "promote"). For this deal, Zaski assumed that beyond the pref, Wildcat would keep 30% of the cash and its limited partners would receive 70%, although promote structures had become a key negotiation point in the difficult economic environment.

Equity cash flow coming from the sale of the property would by distributed somewhat differently. When Wildcat sold the property, Zaski assumed that after-debt proceeds would first repay its limited partners their invested capital (including missed prefs, if any). Then, Wildcat would receive its invested capital. If cash still remained, the limited partners would receive 70%, with Wildcat keeping the remaining 30%.

Wildcat would also earn a 1.5% fee annually, charged against the initial amount of money raised from limited partners. This fee would be taken from the operating cash flow of the property and would not be considered when calculating returns to Wildcat, as this fee compensated the firm for its property management and not for its investment.

The underwriting

Given the assignment of building a financial model for Financial Commons, Zaski began gathering the details she would need to build a 6-year cash flow projection (pro forma). She chose 6 years because Wildcat typically had a hold period of 3 to 5 years, and she would need a forecast of net operating income (NOI) in Year 6 to estimate a projected sale price if the property were held for the full 5 years.

The challenge for Zaski was to make realistic assumptions about the cash flows associated with the property, such as future rent and expenses. As her real estate finance professor often cautioned, "Skilled financial analysts can make a spreadsheet justify anything – so think carefully about your assumptions."

Her plan was to build a benchmark scenario based on her expectations about what would happen. She would then see how sensitive her results were to variations in her assumptions as well as to a few specific adverse scenarios.

Tenant stability was especially important in the recessionary economy; property owners were being hit hard by defaults and vacancies as tenants went bankrupt. The rent roll revealed that Financial Commons' tenants appeared to be a stable and diversified mix, including the largest locally owned accounting firm in Illinois, the headquarters and lead branch of a significant local bank, a trade association for CFAs, an investment advisory services group, a large auto lender, and a national risk and claims management services group. Zaski knew that tenant stability was an especially delicate set of projections to make, as it seemed no one knew when a recovery was coming, and it was difficult to know how these particular tenants would fare in the next few years. However, she did feel confident that the Skokie area would support a steady demand for B+ office space in the long run.

To begin constructing the pro forma, Zaski read through each existing lease carefully. All leases were dated January 1 of various years and would expire on December 31 of the lease expiration year. The leases were either triple net (NNN) or modified gross, and Zaski made a simple table (see Table 4.10) showing lease terms for each tenant. The leases called for annual rent increases of 2%. As owner, Wildcat would be responsible for paying all operating expenses, with the tenants reimbursing Wildcat a fixed amount (per square foot) depending on their lease type, with the modified gross lease tenants providing reimbursements at a lower rate.

Zaski liked that no leases would be expiring in the next 2 years and that no more than 2 tenant leases expired in the same year, reducing the risk of highly concentrated vacancy. From what she had learned, all of the tenants were pleased with management (which Wildcat planned to keep on after the acquisition) and would most likely renew their leases as they came up. Balancing the attractiveness of the local market against the more fundamental weakness in leasing markets generally, Zaski assumed that all existing leases would renew at 2% above the previous year's base rent, contain the same 2% annual rent escalation clauses, and maintain the same level of reimbursable expenses. She further assumed that she would not need to deduct leasing expenses to renew existing tenants.

Zaski also thought it reasonable to assume that Wildcat could lease 1 of the 3 vacant spaces in time to collect rent during the first year of ownership. She further assumed that 1 of the remaining 2 vacant spaces would lease in Year 2, with the final space leasing in Year 3.

Table 4.10 Rent roll of Financial Commons

Tenant	Square Feet	Base Rent Year 1	Annual Rental Adjustment	Remaining Lease Maturity (years)	Lease Type	Reimbursed Expenses (per sq. ft.)
Summer Weill	9,959.00	$241,505.75	2%	7	Modified gross	2
North Shore Bank & Trust Co.	44,280.00	$737,262.00	2%	3	NNN	5
Illinois Institute of CFAs	10,250.00	$214,737.50	2%	6	NNN	5
Beverly & Torres	2,844.00	$66,265.20	2%	7	Modified gross	2
iFinance Dealer Services	6,741.00	$164,480.40	2%	5	Modified gross	2
APQ Consulting	3,612.00	$90,480.60	2%	8	Modified gross	2
Suite 102 (currently vacant)	2,708.67	$56,882.00	0%	NA	NNN	5
Suite 107 (currently vacant)	2,708.67	$56,882.00	0%	NA	NNN	5
Suite 205 (currently vacant)	2,708.67	$56,882.00	0%	NA	NNN	5

96 Equity partnerships

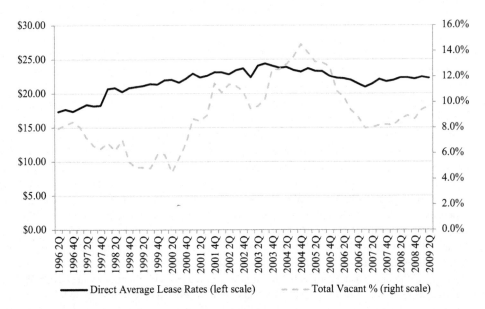

Figure 4.8 Near North leasing and vacancy rates
Source: CoStar Group, Inc.

To determine the rental rate for the new leases, Zaski examined recent vacancy and leasing trends for the area (Figure 4.8). The average rental rate in the area was $21.80 per square foot at the end of the second quarter of 2009, but Zaski felt that because of the market downturn, $21 would be both prudent and realistic for the next couple years and that these new leases would not have annual rent increases. Each new tenant would be found through the help of a leasing broker, who would charge 2% of the first year's gross rent as commission.

As was typical in the industry, Zaski assumed that an additional 3.5% of realized rental income (potential gross income less vacancy) would be allocated for credit losses.

She saw from the property's financial statements that total operating expenses came to roughly 61% of realized rental revenue, so she used this ratio to forecast operating expenses in Year 1. Zaski assumed that these expenses were fixed – that is, independent of the level of occupancy – and that they would grow at 2.5% a year.

With an exit in 3 to 5 years, it was likely that Wildcat would not need to make any major capital improvements to the property, so Zaski did not include these expenses in her benchmark scenario.

Zaski also needed to plan for Wildcat's exit. She estimated the sales price of Financial Commons by applying an exit cap rate to the next year's forecasted NOI. For instance, a sale in Year 5 would be forecasted at a price equal to Year 6 NOI divided by an exit cap rate. Based on the experience of Tripp and the rest of the Wildcat team, Zaski assumed an 8.4% cap rate on exit. Although dismayed by the dearth of hard data and the complete lack of comparable office transactions in the submarket for more than a year (Table 4.11), Zaski saw that Wildcat could back into an 8.4% cap rate by assuming a sale at roughly $140 per square foot in 3 years, which Tripp believed was at the conservative end of what the region would support after credit markets stabilized.

Table 4.11 Comparable office sales

	Closing Date	Square Feet	Sales Price	Price per Sq. Ft.	Cap Rate at Sale	Occupancy at Sale
1033 University Place, Evanston	Jun-08	92,520	$19,200,000	$208		82%
8255 N. Central Park Ave., Skokie	Mar-08	283,008	$11,000,000	$39		
5700 Old Orchard Rd., Skokie	Jan-08	34,365	$5,000,000	$145		66%
5200–5202 Old Orchard Rd., Skokie	Dec-07	350,559	$64,000,000	$183	5.0%	76%

Source: Real Capital Analytics, Inc.

Table 4.12 Cash flow pro forma for Financial Commons

	2009 Year 0	2010 Year 1	2011 Year 2	2012 Year 3	2013 Year 4
Potential gross income		$1,685,377.45	$1,715,672.08	$1,746,572.60	$1,778,091.13
Vacancy		$113,764.00	$56,882.00		
Credit loss		$55,006.47	$58,057.65	$61,130.04	$62,233.19
Effective gross income		$1,516,606.98	$1,600,732.43	$1,685,442.56	$1,715,857.94
Expense reimbursements		$332,505.33	$346,048.67	$359,592.00	$359,592.00
Total operating revenue		$1,849,112.31	$1,946,781.09	$2,045,034.56	$2,075,449.94
Operating expenses		$958,684.20	$982,651.31	$1,007,217.59	$1,032,398.03
Net operating income		$890,428.11	$964,129.78	$1,037,816.97	$1,043,051.91
Capital expenditures					
Leasing commissions		$1,137.64	$1,137.64	$1,137.64	
Management fee		$51,870.00	$51,870.00	$51,870.00	
Reversion sale price				$12,417,284.65	
Property cash flow from operations		$837,420.47	$911,122.14	$984,809.33	
Total property cash flow	($10,400,000.00)	$837,420.47	$911,122.14	$13,402,093.98	

The deal

Wildcat was presented with the opportunity to buy the property for $10,400,000. This price represented a 28% discount from the current owner's purchase price and was more than 13% below the currently outstanding debt.

Zaski set to work to determine if Financial Commons was a worthwhile investment for Wildcat in its real estate acquisitions–focused business. She began by developing a cash flow pro forma for Financial Commons assuming that Wildcat would sell the property after a 3-year hold using her benchmark assumptions (Table 4.12). Having completed

98 Equity partnerships

Table 4.13 Equity waterfall template for Financial Commons

	2009	2010	2011	2012
	Year 0	Year 1	Year 2	Year 3
Equity-Level Cash Flows:				
Equity-level operational before-tax cash flow				
Equity-level reversion before-tax cash flow				
Equity-level before-tax cash flow				
Investor Equity Capital Account:				
Beginning equity investment balance				
Annual preferred investment				
Preferred return earned				
Preferred return paid				
Accrued but unpaid preferred return				
Ending equity investment balance				
Wildcat Equity Capital Account:				
Beginning equity investment balance				
Annual subordinated investment				
Ending equity investment balance				
Operational Cash Flow:				
Investor-level cash flows				
Wildcat cash flows (excluding management fee)				
Reversion Allocations:				
Investor return of equity (with preference)				
Wildcat return of equity				
Investor additional proceeds				
Wildcat additional proceeds				
Reversion Cash Flow:				
Investor-level cash flows				
Wildcat-level cash flows				
Total Cash Flow to Equity:				
Investor-level cash flows				
Wildcat-level cash flows				

that analysis, Zaski next had to consider how the terms of the mortgage financing and the terms of the partnership agreement would interact to determine the expected returns to Wildcat and its limited partner investors. She developed a template for modeling the equity before-tax cash flows that each investor group would receive (Table 4.13) and began her work.

As careful as she had been with the assumptions in her benchmark analysis, Zaski knew the future had a way of diverging from expectations, so she began to review and test her underlying assumptions to determine which were most important in determining the ultimate return received by Wildcat and its investors.

Notes

1 In practice, when negative equity cash flow is *expected* during the holding period – as it is in this particular example – the additional capital is often raised at the outset of the investment so that a later capital call is not needed. Capital calls made after property acquisition, in practice, therefore typically reflect unexpectedly poor property performance.
2 A different partnership agreement might have called for the repayment of missed preferred returns before any cash flow goes through a promote. To simplify the discussion, we do not consider this possibility.

5 The taxation of property investment

Introduction

Chapter 5 describes the taxation often faced by investors in property and, in so doing, demonstrates how taxes reduce the ultimate return an investor will earn. Note that the taxation of investment property, like most aspects of taxation, is complicated. In this chapter, we simplify the rules to illustrate the key principals of real estate taxation for a US-based taxpayer. In the accompanying case study, you will understand how taxation affects the returns to investing in property in the United Kingdom. In the end, the objective of the chapter is not to have you become an expert in taxation, but to realize that actions that you take as a property investor do impact the amount you will owe in taxes, and thus, your after-tax rate of return on the investment. In particular, the amount you borrow influences the amount you pay in interest, which affects your taxable income. The amount you invest in capital expenditures affects your cost basis and your allowable depreciation expenses, which in turn, affect taxes due when you sell the investment as well as along the way.

In this chapter, we will additionally assume that there is only a single equity investor in a given property. We make this assumption for simplicity. Recall that in Chapter 4 we considered that there may be an equity partnership involving 2 or more equity investors and demonstrated how a partnership agreement would allow the equity level cash flow to be split between equity investors. The cash flow that flowed through the waterfall was before-tax cash flow. The reason that real estate private equity partnerships calculate cash flow splits on before-tax cash flow is because it is conceivable that some partners face different tax rates than others. Therefore, an investor who is a limited partner (LP) in a partnership would receive distributions from the partnership and then have to calculate the investor-specific taxes due.

Income taxes during the holding period

We will consider the problem of calculating taxes due during the holding period separately from the calculation of taxes due at the time of reversion because there are different tax rules that apply to these different sources of cash flow. Simply stated, income generated during the holding period is subject to **income taxes**, whereas income from the ultimate sale of the property is subject to capital gains taxation.

To calculate the income taxes due, we follow the steps outlined in Table 5.1.

As indicated in Table 5.1, the calculation of income taxes required to be paid by a property investor requires a calculation of taxable income. In turn, taxable income is equal to property NOI less both interest and depreciation. Note that the calculation of taxable income starts from NOI and not total property cash flow.

Table 5.1 Calculation of income taxes

Net operating income (NOI)
Less interest paid
Less depreciation
= Taxable income
× Investor's tax rate (may vary according to investor)
= Income taxes due

Table 5.2 Pro forma with debt service cash flows

	Year 0	Year 1	Year 2	Year 3	Year 4	Year 5
Net operating income (NOI)		$184,680	$405,240	$337,280	$481,460	$477,822
Building improvements		$0	$0	$80,000	$0	$0
TIs		$60,000	$0	$60,000	$0	$0
Leasing commissions		$165,750	$0	$180,000	$0	$0
Holding period cash flow		−$41,070	$405,240	$17,280	$481,460	$477,822
Reversion cash flow						
Sales price						$5,926,615
Sales commission						$177,798.46
Total property cash flow	−$4,600,000	−$41,070	$405,240	$17,280	$481,460	$6,226,639
Interest paid		$126,169	$123,185	$120,071	$116,823	$113,434
Principal repayment		$68,857	$71,841	$74,954	$78,203	$2,706,146
Total debt service	−$3,000,000	$195,026	$195,026	$195,026	$195,026	$2,819,580
Total cash flow to equity	−$1,600,000	−$236,096	$210,214	−$177,746	$286,434	$3,407,059

Interest paid

In Chapter 3, we saw how we can use our rules of mortgage amortization to separate our monthly mortgage payment into interest and principal. In Table 5.2, we expand upon our property pro forma from Chapter 3 in order to highlight this distinction. Importantly for our income tax calculation, interest and not principal repayment is tax deductible. As we calculated before, the interest paid in the first year of property ownership is $126,169. This amount falls gradually over time as the loan is slowly repaid.

Depreciation

The physical structure of an investment property deteriorates over its lifetime. This is what motivates the capital expenditures described in Chapter 2. However, this physical or economic depreciation is not what influences an investor's tax liability. Rather, the **depreciation** that is relevant for an investor's income tax calculation is the depreciation allowed by the tax code. Somewhat simplifying the actual rules for US-based taxpayers, we will assume that the structure component of a property investment can be depreciated over 30 years if the property is a residential building. Other investment property structures are depreciated over 40 years. This means that to calculate the depreciation allowable for tax purposes, we need to know the fraction of our cost basis (purchase price) that is for the land as opposed to the structure built on the land. The land component is not depreciable for tax purposes.

102 The taxation of property investment

In addition to depreciating the physical structure of an investment property, an investor can also depreciate for tax purposes building improvements and leasing costs. In practice, each building improvement is assigned a specific depreciation horizon, while leasing costs depreciate over the horizon of the relevant lease. To simplify our illustrations by eliminating the need to explicitly describe each building improvement and its unique depreciation schedule, I will assume that the building improvements in our example are able to be depreciated over 3 years. Leasing costs – both TIs and leasing commissions – will be depreciated over the life of the lease.

Continuing with our example, suppose that it can be determined that 30% of the value of the industrial building is in the land. Thus, the industrial structure can be depreciated in an amount of $80,500 (70% × $4,600,000 ÷ 40) in Years 1–40. Tenants in the building have been assumed to sign 10-year leases. Therefore, the $60,000 TI in Year 1 will generate $6000 in depreciation in Years 1–10. Analogously, the $165,750 leasing commission in Year 1 will generate $16,575 of depreciation in Years 1–10. Building improvements of $80,000 in Year 3 will generate $26,667 of depreciation in Years 3–5.

Income taxes due

Income taxes owed on the taxable income during the holding period is calculated by multiplying the investor's marginal income tax rate by taxable income. In this example, I assume a combined federal and state marginal tax rate of 40%, but of course this rate would vary across investor.

Because vacancy is high in the first year of operation, taxable income is negative. When faced with negative taxable income, investors do not receive a tax refund. Rather, this negative taxable income can be rolled forward as a tax loss carry forward to be used to offset future positive taxable income. As shown in Table 5.3, this implies that taxes due in Year 1 are $0. Taxable income in Year 2 is partially offset by this tax loss carry forward, so that income taxes due in Year 2 are equal to $53,766 = 40% × ($178,980 – $44,564). By assuming our investor makes use of a tax loss carry forward, we are assuming that the investor has no other source of passive income against which the loss can be offset. By contrast, if our investor had another investment property generating positive taxable income, then the taxable loss on this property could offset the taxable income on the other, thus allowing the investor to immediately realize the tax savings at the time the losses occur.[1]

Table 5.3 Calculation of income taxes due

	Year 1	Year 2	Year 3	Year 4	Year 5
Net operating income (NOI)	$184,680	$405,240	$337,280	$481,460	$477,822
Interest paid	$126,169	$123,185	$120,071	$116,823	$113,434
Depreciation of					
Structure	$80,500	$80,500	$80,500	$80,500	$80,500
Building improvements	$0	$0	$26,667	$26,667	$26,667
Tenant improvement allowances	$6000	$6000	$12,000	$12,000	$12,000
Leasing commissions	$16,575	$16,575	$34,575	$34,575	$34,575
Taxable income	-$44,564	$178,980	$63,467	$210,895	$210,646
Tax loss carry forward	$44,564	$0	$0	$0	$0
Income taxes due	$0	$53,766	$25,387	$84,358	$84,258

Capital gains taxes due at reversion

Assuming (as we hope) that our investment property sells for a price higher than we originally paid, we will have to pay additional taxes on this capital gain. The overall capital gain on the investment property is the difference between the sales price (net of any sales expenses) and the book value of the property. The book value, in turn, is the purchase price plus any capital expenditures less any depreciation during the holding period. In our industrial building example, the total capital gain is $1,270,442, which is the difference between the net sales price of $5,748,817 and the book value of the property, which at the end of Year 5 is $4,478,375.

Typically, capital gains taxes due would be calculated by multiplying a capital gains tax rate by the gain, but for US taxpayers recognizing a gain on their investment property the actual calculation is more complex. This is because the US tax code distinguishes between 2 separate events that cause a property sales price to exceed its book value. As shown graphically in Figure 5.1, it is possible to deconstruct the total capital gain (sale price less adjusted basis) into 2 component parts. The first is called price appreciation and the second is called depreciation recapture. This distinction is important because the tax rates paid on each component are different.

Taxes on depreciation recapture

Depreciation recapture is the mechanism by which the government partially recovers the depreciation tax benefit given to a property investor when the investor has sold an asset that has *appreciated* in value. The calculation of depreciation recapture taxes is simply equal to 25% of the total amount of depreciation taken during the holding period. In our example industrial property, total depreciation would reflect (1) the depreciation of the structure for

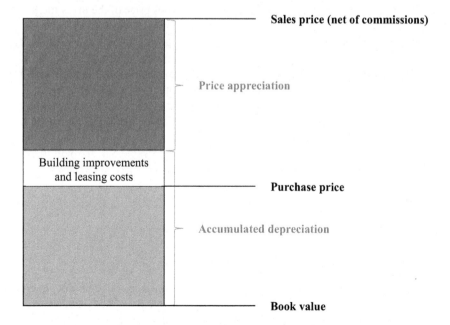

Figure 5.1 A graphical representation of capital gains taxation

104 The taxation of property investment

Table 5.4 Capital gains taxation of investment property

	Year 0	Year 1	Year 2	Year 3	Year 4	Year 5
Cost basis of property	$4,600,000	$4,722,675	$4,619,600	$4,785,858	$4,632,117	$4,478,375
Net sales price						$5,748,817
Accumulated depreciation						$667,375
Price appreciation						$603,067
Capital gains taxes due						$287,457

the first 5 years out of the 40 years over which it depreciates according to the tax code, (2) the partial depreciation of the leasing commissions, (3) the partial depreciation of tenant improvement allowances, and (4) the complete depreciation of the building improvements. In our example, the investor took total depreciation deductions of $667,375. Depreciation recapture taxes of $166,844 = 25% × $667,375 would therefore be due upon sale. Despite the recapture of depreciation benefits upon the sale of property, the ability to depreciate investment property is still beneficial to the investor. First, it allows the postponement of taxes until the property is sold. Second, the tax rate on depreciation recapture is generally lower than the investor's tax rate on income, which reflects the initial tax benefit of depreciation. Thus, the ability to depreciate investment property that is appreciating not only delays but reduces an investor's total tax obligation.

Taxes on price appreciation

Price appreciation is defined as the part of the capital gain that is not caused by depreciation. That is, **price appreciation** is the difference between the sales price of the investment property after subtracting any sales commissions and the total amount of capital invested into the property, which combines the initial purchase and any capital expenditures. In our industrial property, the total capital gain was $1,270,442 and there was total depreciation of $667,375. Therefore, price appreciation was $603,067 = $1,270,442 − $667,375. Our investor owes price appreciation taxes in an amount equal to the price appreciation gains times the investor's marginal tax rate on capital gains. If we assume our investor faces the highest marginal capital gains tax rate of 20%, then the investor owes 20% × $603,067 = $120,613 in taxes resulting from price appreciation.

The inputs to the capital gains tax calculation are shown in Table 5.4. In total, therefore, capital gains taxes of $287,457 (both depreciation recapture and price appreciation) are due upon sale.

Putting it all together

After-tax cash flow to the equity investor

We have now calculated the inputs necessary to determine the after-tax cash flow to equity. Beginning with the total before-tax cash flow to equity we calculated in Chapter 3, we then subtract (1) income taxes and (2) capital gains taxes. For our example industrial property, after-tax cash flows can be calculated as shown in Table 5.5.

Table 5.5 Equity after-tax cash flow

	Year 0	Year 1	Year 2	Year 3	Year 4	Year 5
Total property cash flow	−$4,600,000	−$41,070	$405,240	$17,280	$481,460	$6,226,639
Interest paid		$126,169	$123,185	$120,071	$116,823	$113,434
Principal repayment		$68,857	$71,841	$74,954	$78,203	$2,706,146
Total debt service	−$3,000,000	$195,026	$195,026	$195,026	$195,026	$2,819,580
Total cash flow to equity	−$1,600,000	−$236,096	$210,214	−$177,746	$286,434	$3,407,059
Income taxes due		$0	$53,766	$25,387	$84,358	$84,258
Capital gains taxes due						$287,457
Equity after-tax cash flow	−$1,600,000	−$236,096	$156,448	−$203,132	$202,076	$3,035,343
Property return	9.52%					
Debt return	4.20%					
Equity before-tax return	16.26%					
Equity after-tax return	12.26%					

Calculating the expected return to the equity investor

Given these after-tax cash flows, a potential investor will then calculate an equity-level internal rate of return (IRR) based upon these cash flows. Doing so indicates that the after-tax IRR to the equity investor is forecasted to be 12.26%. This compares to a property level expected return of 9.52% and an equity before-tax rate of return of 16.26%. Note that the forecasted IRR to equity is higher than the expected return on the underlying property. This results from the expected increase in returns arising from leverage being only partially offset by a drag on returns from taxes.

Decision making

So as a potential investor, does the finding that you expect to receive a 12.26% return on your equity investment help you decide whether or not to acquire the property? You may be surprised to realize that the answer is no! As was described in Chapter 2, the appropriate investment decision is made by determining whether or not you can acquire the property at less than its true value. By acquiring the property at less than its estimated value, you have made a positive net present value (NPV) investment. The fact that equity earns a higher rate of return than the underlying property is simply reflecting that, in the presence of leverage, the equity investment is riskier. Earning a higher return is compensation for the additional risk.

Deferral of capital gains taxes

This chapter has discussed the taxation of property investments. In that discussion, we described the capital gains taxation that become due upon sale – price appreciation taxes and depreciation recapture taxes. In this section, we describe 2 methods by which US taxpayers may elect to defer the payment of these taxes due upon sale. The first method is by participating in what is called a 1031 exchange, which is named after the section of the tax code where such exchanges are described. Essentially, a 1031 allows an investor to defer the payment of price appreciation taxes and depreciation recapture taxes if they apply those gains to the purchase of a new, like-kind property. This is why these exchanges are sometimes called like-kind exchanges. The second method to defer capital gains taxes is by reinvesting gains into property located within an "**opportunity zone**." Investors recognizing gains that are then invested into opportunity zones not only defer paying capital gains taxes, but the taxes owed are reduced if the investment lasts for at least 5 years, and capital gains taxes on the new investment in the zone are completely eliminated if the investment lasts for at least 10 years.

Requirements of a 1031 exchange

There are many rules and restrictions in place for the sale and purchase of investment property to get the tax benefits associated with the deferral of capital gains taxes. In this section, we outline the main restrictions.

Qualified use

The properties involved in the exchange must be held for investment or used in the investor's trade or business. In particular, the qualified use test excludes an individual's primary residence.

Property must be like-kind

In our example that follows, an investor exchanges an industrial building for a retail center. Essentially any type of investment property would satisfy the conditions that the new investment was of "like-kind" to the original. Investors can exchange multifamily properties, shopping centers, industrial buildings, office towers, farmland, or other types. However, there are some exceptions. One cannot exchange an investment in a property for shares in a real estate company, for the former is an equity interest in property while the latter is a security interest. Exchanging domestic property for other domestic property is allowed. So is exchanging foreign property for foreign property. However, you are not allowed to exchange domestic property for foreign property. Land and structures can be exchanged for either land or structures, but land alone cannot be exchanged for only structures.

Timing

For a sale and purchase of investment property to qualify for a 1031 exchange, the new property being acquired must be identified within 45 days and the exchange must be completed within 180 days after the title transfer of the exchanged property.

Use of a qualified intermediary

The exchange will be disqualified if the investor ever has the right to access, control, or receive, or could have received, the sale proceeds of the original investment property. For this reason, an investor must use a 1031 "Exchange Accommodator" or "Exchange Facilitator" to hold onto the sale proceeds during the time after the sale of the initial property and the acquisition of the new property.

Neither the equity investment nor the mortgage debt may decline

Suppose our investor of our 2-tenant industrial property wished to conduct a like-kind exchange. Our investor sold his industrial property and received an equity payout of

$3,124,263, which represents the property sales price of $5,926,615 less $177,798 in commissions and the lump sum repayment of the outstanding mortgage balance of $2,624,554. For a 1031 exchange to be free from tax, our investor would have to put at least $3,124,263 of equity into the new property. At the same time, the investor must not receive any reduction in mortgage balance, which at the time of sale was $2,624,554. When properties are appreciating in value, an investor doing a 1031 exchange will generally be exchanging into more expensive property, so that the entire equity investment can be exchanged and the mortgage balance not be reduced. Any equity or debt reduction received by the investor as part of the exchange would be immediately taxable.

The mechanics of a 1031 exchange

We will illustrate the mechanics of a **1031 exchange** by continuing our example of an investment into a 2-tenant industrial property. As was illustrated in Table 5.5, our investor was subject to capital gains taxes of $287,457, which was based on a taxable gain of $1,270,442. If instead the investor completes a tax deferred exchange by satisfying all of the requirements listed previously, the investor will not have to immediately pay $287,457 in capital gains taxes. However, by conducting the exchange, the gain of $1,270,442 is transferred to the new property, thus lowering the cost basis of the new acquisition.

The reduction in the cost basis of the new property has 2 implications. First, it will increase the income taxes the investor will pay during the holding period of the second property relative to what would be paid on the second property had an exchange not been completed. This is because the lower basis implies lower depreciation allowances. In particular, suppose the new investment property is a retail shopping plaza and thus also allows depreciation of its structure over 40 years. If 30% of the new retail shopping plaza is attributed to land, then because the investor has undertaken the exchange, depreciation on the new property will be 70% × $1,270,442 ÷ 40 = $22,233 lower each year than it would have been absent the exchange. As our investor pays 40% of income in taxes, the exchange has caused income taxes to rise by 40% × $22,233 = $8893 each year the retail center is owned.

The second implication of rolling the gains from the industrial building into the shopping center is that capital gains taxes upon selling the retail building will be higher than they would be were it not for the exchange. To see why, realize that price appreciation gains are unaffected by the exchange. In our example, the investor had achieved a price appreciation gain of $603,067 from the industrial building. Taxes on those price appreciation gains are simply deferred until the retail shopping center is sold.[2] This is because the exchange has no impact on either the sales price of the shopping center or the total amount of money the investor has put into both properties in the form of capital expenditures. However, the investor also transferred $667,375 of depreciation recapture to the new property. Each year, however, the relative depreciation of the retail building is $22,233 less because of the exchange. Therefore, the additional depreciation recapture subject to taxation upon sale of the retail building is $667,375 − ($22,233 × holding period of retail building).

> Example 5.1: Calculate the financial tradeoffs of completing this exchange if the investor holds the retail shopping center for 5 years.

The benefit of this transaction is that the investor does not pay $287,457 in capital gains taxes at the time of the property exchange. The cost of the exchange is twofold. First, the investor will pay $8893 in additional income taxes each of the 5 years of owning the shopping center.

108 The taxation of property investment

Second, the investor will owe additional capital gains taxes upon selling the retail building. These additional capital gains taxes are the sum of the (unaffected) price appreciation and the (affected) depreciation recapture taxes. The price appreciation taxes will be 20% × 603,067 = $120,613 higher because of the exchange. The depreciation recapture taxes will be 25% × ($667,375 − [$22,233 × 5]) = $139,053 higher than they otherwise would be upon selling the retail building due to the investor having done an exchange. In total, the exchange causes the investor to pay $259,666 = $120,613 + $139,053 more in capital gains taxes upon the sale of the retail building. These cash flow implications of the exchange are shown in Table 5.6.

We can calculate the IRR of the cash flows in the last column of Table 5.6 to be 1.21%. Although this sounds like a low return, realize that the pattern of cash flows being examined have the opposite signs of a typical investment. In particular, the immediate cash flow is positive and the subsequent cash flows are negative. Thus, completing an exchange is like *borrowing* money rather than investing. Therefore, the results of this example imply that completing the exchange is economically equivalent to borrowing money (from the government) at a 1.21% interest rate. If this is lower than the investor's borrowing rate, then doing the 1031 exchange would be a positive NPV choice.[3]

In this example, we made a number of assumptions before deriving the implicit cost of doing an exchange. We can generalize these findings by calculating the implicit borrowing cost of completing an exchange as we change some of the underlying assumptions. Table 5.7 reports the implicit borrowing rate by an investor completing a 1031 exchange when we vary the investors marginal income tax rate and the percent of the new investment attributable to land.

As indicated in Table 5.7, lower income tax rates increase the benefits of an exchange. This is intuitive because higher income taxes (from lower depreciation) are one explicit cost

Table 5.6 Cash flow implications of an exchange

Years Relative to the Exchange	Increased Income Tax	Increased Capital Gains Tax	Total Cash Flow
0			$287,457.13
1	−$8893.09	$0.00	−$8893.09
2	−$8893.09	$0.00	−$8893.09
3	−$8893.09	$0.00	−$8893.09
4	−$8893.09	$0.00	−$8893.09
5	−$8893.09	−$259,666.21	−$268,559.31

Table 5.7 Income tax and land sensitivity

		Percent Attributable to Land			
		0%	15%	30%	100%
Income tax rate	20%	−0.58%	−0.49%	−0.40%	0.00%
	25%	0.00%	0.00%	0.00%	0.00%
	30%	0.58%	0.49%	0.40%	0.00%
	35%	1.17%	0.98%	0.80%	0.00%
	40%	1.75%	1.48%	1.21%	0.00%
	45%	2.34%	1.97%	1.61%	0.00%

of completing the exchange. Note that at an income tax rate of 25%, an exchange is like borrowing from the government at a 0% interest rate. This is because at a 25% income tax rate, the higher taxes incurred by doing the exchange are exactly the same as the additional depreciation recapture taxes paid upon sale since depreciation recapture also faces a 25% marginal rate. Therefore, at a 25% marginal income tax rate, an exchange is economically equivalent to a deferral of taxes at a 0% interest rate. Lower marginal income tax rates would actually generate a positive return from the exchange. In such an instance, the government would be paying you to complete the exchange.

Table 5.7 also indicates that for income tax rates above the recapture rate of 25%, the more the purchase price of the new property is attributable to land, the greater the benefit (i.e. the lower the implicit borrowing cost) to the exchange. This makes sense, too, because land is not depreciable. Therefore, the cost of lost depreciation falls when the new property's land value rises. In the extreme, when an investor exchanges into raw land, it is economically equivalent to borrowing from the government at a 0% interest rate. Intuitively, this is because buying land has no depreciation benefit and therefore, no income tax penalty from the exchange is realized.

Table 5.8 explores the economic tradeoffs of an exchange as the holding period of the investor in the second property and the capital gains tax rates vary. In this table, it is assumed that the investor's marginal income tax rate remains at 40% and the fraction of the exchanged property's value attributable to the land remains at 30%. As illustrated in the top panel, regardless of the holding period, the marginal benefit of an exchange rises with higher price appreciation taxes. This is because price appreciation taxes are simply deferred and a higher tax rate on such gains implies more dollars of taxes are simply being deferred. In the bottom panel, the same relationship is found for depreciation recapture taxes. The magnitude of the relationship between marginal tax rates and implicit borrowing costs is greater for depreciation recapture than it was for price appreciation. This occurs because an exchange not only defers depreciation recapture, but the amount of depreciation recapture taxes paid upon the sale of the second property falls as less depreciation is taken during the holding period of the second property.

Table 5.8 Capital gains and holding period sensitivity

		\multicolumn{4}{c}{Holding period}
		5	10	15	20
Price appreciation tax rate	20%	1.21%	1.27%	1.33%	1.41%
	25%	1.09%	1.14%	1.19%	1.25%
	30%	0.99%	1.03%	1.08%	1.12%
	35%	0.91%	0.94%	0.98%	1.02%
	40%	0.84%	0.87%	0.90%	0.93%
	45%	0.78%	0.81%	0.83%	0.86%
		\multicolumn{4}{c}{Holding period}
		5	10	15	20
Depreciation recapture tax rate	20%	1.81%	1.89%	1.98%	2.07%
	25%	1.21%	1.27%	1.33%	1.41%
	30%	0.72%	0.76%	0.81%	0.86%
	35%	0.33%	0.35%	0.37%	0.40%
	40%	0.00%	0.00%	0.00%	0.00%
	45%	−0.28%	−0.30%	−0.32%	−0.34%

Table 5.8 also indicates that longer holding periods of the second property increase the implicit borrowing cost of an exchange. This is because the marginal cost of an exchange is higher income taxes, and a longer holding period increases the time over which these higher income taxes are paid.

Opportunity zones

In 2017, US tax legislation created "opportunity zones," which are specific geographic locations in which real estate investment would receive preferential tax treatment. Like 1031 exchanges, opportunity zones allow an investor to defer capital gains taxation by rolling a capital gain from an existing property investment into a new investment located within a certain zone.

In some ways, opportunity zone investing is clearly superior to 1031 exchanges. First, to benefit from the deferral of capital gains taxes, an investor only needs to invest the gain and is not subject to the requirement that the equity and debt investment into the new property be at least as large as the old. Second, the source of the gain need not be like-kind. That is, an investor can realize a gain on a stock sale, invest in real estate in an opportunity zone, and still realize the tax benefits. Third, opportunity zone investors receive a 10% reduction in their deferred capital gains taxes owed if the investment lasts for at least 5 years. Finally, and perhaps most significantly, capital gains on the new opportunity zone investment are completely eliminated if the investment is held for at least 10 years.

On the other hand, deferring capital gains taxes through opportunity zone investment may not be appropriate for all property investors. First, the deferral of capital gains taxes only lasts until December 31, 2026. Thus, as time passes, the deferral becomes less valuable both because the time of deferral shortens and because the investment becomes less likely to achieve the 5-year investment horizon necessary for the tax reduction. Second, opportunity zone investment must satisfy a "substantial improvement" requirement that implicitly means that to take advantage of the tax benefits of opportunity zones, investors will not be able to invest in cash-flowing assets generating steady income but rather must invest in development or redevelopment projects that have substantially more risk.

Because opportunity zone investing is still new at the time of this writing, it remains to be seen how important such an approach will become for most property investors seeking to defer capital gains taxes.

The taxation of property investment: a landlord's certainty

In this case study, you are put in the shoes of Heather Wilson, a management consultant living in London. Wilson is finalizing the acquisition of her first investment property, a flat in the desirable Kensington and Chelsea area. In completing this case, you will be able to:

1   Calculate the taxes that must be paid when undertaking an investment in property.
2   Understand how taxes influence the rate of return earned by a property investor.
3   Determine the relationship between financing choice and after-tax returns.
4   Make sense of the theory of optimal capital structure in a real estate context.

A landlord's certainty

December 16, 2015, was an exciting day for 32-year-old Heather Wilson. After years of painstaking savings, she had finally reached an agreement to purchase her first buy-to-let property, a 1-bedroom flat in London's sought-after Kensington and Chelsea neighborhood. She looked forward to a lifetime of building wealth through property investments. Of course, some of the income the property would generate would be owed to Her Majesty's Revenue and Customs (HMRC). But such was the nature of life. Unfortunately, the tax laws had recently become less favorable for property investors, but Wilson expected to negotiate a lower purchase price as a result and so she felt confident that her investment remained solid.

Heather Wilson

Wilson had always been intrigued with real estate. Living in London her entire life, she had experienced a period of time that witnessed tremendous swings in real estate prices (Figure 5.2). After prices collapsed in 2008 and 2009, she took what little savings she had amassed and bought a small 2-bedroom flat in the Kingston upon Thames area. The rebound in property prices lifted the value of her home over the following years, so when she remortgaged the flat she was able to extract substantial equity. When combined with diligent savings out of her management consultant paycheck, this meant that now Wilson had approximately £300,000 she could invest elsewhere. She viewed real estate as a way to build lifetime wealth, and so she began looking for buy-to-lets. Because she was relatively risk averse, she looked for investment property closer to central London, which ultimately led her to discover the property in the Royal Borough of Kensington and Chelsea.

The Property

Wilson's buy-to-let property was located on Cromwell Road, a few minutes' walk from the Earl's Court and Gloucester Road tube stations (Figure 5.3). It was a 1-bedroom flat located at the back of the property, away from the noise of Cromwell Road, with south-facing views

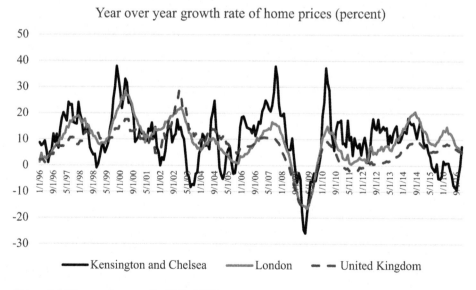

Figure 5.2 House price growth, 1996–2016
Source: United Kingdom Office for National Statistics House Price Index.

112 *The taxation of property investment*

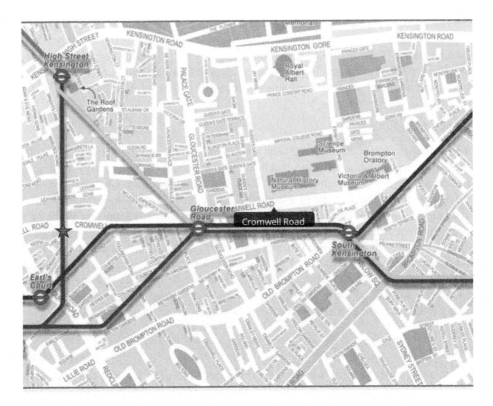

Figure 5.3 Property location

Note: Location of property marked by star

Source: LondonTown, "Cromwell Road Location Map," accessed February 15, 2019, http://www.londontown.com/LondonStreets/cromwell_road_909.html.

and approximately 43.9 square meters in size (Figure 5.4). From the property, her tenant would be able to be in Central London in 10 minutes via the Piccadilly line, and in East London in less than 30 minutes via the Circle line. The property was also within walking distance of London's internationally renowned Victoria and Albert Museum, the Royal Albert Hall, and the lovely Kensington Gardens. The property was listed for sale at £565,000, and so, as a benchmark, Wilson assumed this would be her acquisition price. However, she was still in negotiations with the seller, and in light of the new tax rules Wilson thought she might be able to achieve a lower price.

With such a great location, the property was expected to be very easy to let. In fact, the property already had a tenant paying £425 per week, implying a gross rental yield of 3.9%.[4] Although this yield was low, it was actually higher than the average yields in the Kensington and Chelsea area. The neighborhood was among the city's most attractive, and therefore it was not unusual that yields in the area were among the lowest in London (Figure 5.5). By comparison, the 3.9% yield on Wilson's property looked rather generous. Of course, there would be expenses associated with property ownership, and with a few additional assumptions (Table 5.9) Wilson was able to calculate that the property would deliver a net rental yield of approximately 3.2%.[5]

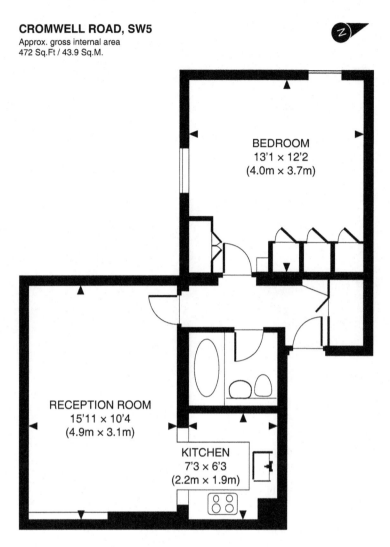

Figure 5.4 Property floor plan

Of course, Wilson expected that the investment would deliver returns in excess of its yield due to price appreciation. After reviewing historical data (Figure 5.6), she assumed that the annual increase in the rent she could collect would be 3% per year, which was the approximate average growth rate of rents over the previous 10 years. Property expenses would probably grow at a similar pace. Looking at historical property prices (Figure 5.2), she calculated that the average growth rate of property prices in the Kensington and Chelsea neighborhood was 10%. This seemed like a very high number, especially since she knew that home prices had been falling recently. Looking only at the most recent 5 years, house prices in the neighborhood had increased at an annual rate of 7.3%. To remain somewhat conservative, Wilson thought it reasonable to assume that the sales price of her investment property would rise by 6% per year. From this, Wilson was able to calculate a forecast of the cash flows her property

114 *The taxation of property investment*

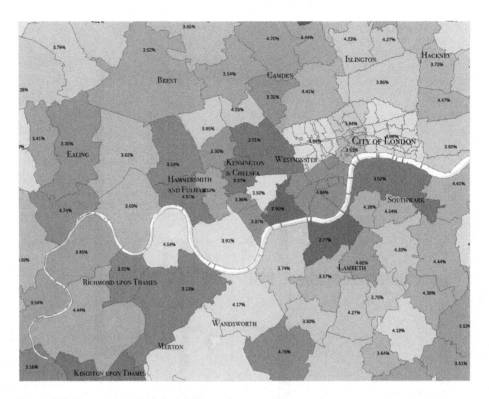

Figure 5.5 Gross rental yields in London and surrounding areas

Table 5.9 Costs of investment property ownership

Vacancy allowance	1 week/year
Property management (as fraction of gross rental revenue)	8%
Monthly maintenance expense	£100
Annual insurance premium	£675
Capital expenditures as a share of net operating income (NOI)	10%

investment would deliver to her over the coming 10 years, from which she calculated a total return on the investment of approximately 8.1% (Table 5.10). At this point Wilson paused, realizing that to justify a return in a spreadsheet only required skill in financial analysis. For those returns to be realized, she needed to be comfortable with her assumptions.

Taxation of investment property

Although she had benefited from good investment timing on her own home, Wilson was not quite as lucky with her Cromwell Road investment. Because of the persistently rapid rise of property prices, the UK government had recently undertaken a number of policy changes aimed at cooling off the very hot buy-to-let market. These policy changes impacted the taxes that Wilson would face as a landlord.

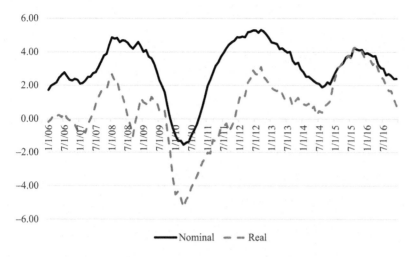

Figure 5.6 Annual growth rate of London rental rates (percentage points)

Source: United Kingdom Office for National Statistics Experimental Index of Private Housing Rental Prices, and consumer price inflation data. The index calculates changes in the rents for both new and ongoing tenancies.

Income tax and the tax deductibility of financing costs

Taxable income in the United Kingdom is equal to earned income less a personal allowance. As a property investor, Wilson's earned income would be the sum of her property's taxable income and her salary, which was currently £135,000 per year.[6] The personal allowance was £11,000, but for taxpayers with earned income in excess of £100,000 the allowance began to be phased out.[7] In particular, the personal allowance was reduced by £1 for every £2 of earned income in excess of £100,000. Once earned income exceeded £123,700, the personal allowance was £0. Individuals with taxable income up to £32,000 paid basic rates of 20%. Those with income between £32,000 and £150,000 were taxed at higher rates (40%). Any taxable income in excess of £150,000 faced an additional rate of 45% (Table 5.11). Thus, Wilson's salary meant she paid higher rates, and soon she would be an additional-rate payer.

Some expenses on her investment property were deductible from gross rental income, for example:[8]

- Letting agents' fees[9]
- Legal fees for lets of a year or less, or for renewing a lease for less than 50 years
- Accountants' fees
- Buildings and contents insurance
- Interest on property loans
- Maintenance and repairs to the property (but not improvements)
- Utility bills, like gas, water, and electricity
- Council tax
- Contracted services such as cleaning or gardening
- Other direct costs of letting the property, like phone calls, stationery, and advertising

Table 5.10 Property pro forma (in £)

Property Pro Forma

	2015	2016	2017	2018	2019	2020	2021	2022	2023	2024	2025	2026
Rent		£22,100.00	£22,763.00	£23,445.89	£24,149.27	£24,873.74	£25,619.96	£26,388.56	£27,180.21	£27,995.62	£28,835.49	£29,700.55
Vacancy		£425.00	£437.75	£450.88	£464.41	£478.34	£492.69	£507.47	£5 22.70	£538.38	£554.53	£571.16
Total revenue		£21,675.00	£22,325.25	£22,995.01	£23,684.86	£24,395.40	£25,127.27	£25,881.08	£26,657.52	£27,457.24	£28,280.96	£29,129.39
Property management		£1,734.00	£1,786.02	£1,839.60	£1,894.79	£1,951.63	£2,010.18	£2,070.49	£2,132.60	£2,196.58	£2,262.48	£2,330.35
Maintenance		£1,200.00	£1,236.00	£1,273.08	£1,311.27	£1,350.61	£1,391.13	£1,432.86	£1,475.85	£1,520.12	£1,565.73	£1,612.70
Insurance		£675.00	£695.25	£716.11	£737.59	£759.72	£782.51	£805.99	£830.16	£855.07	£880.72	£907.14
Total expenses		£3,609.00	£3,717.27	£3,828.79	£3,943.65	£4,061.96	£4,183.82	£4,309.33	£4,438.61	£4,571.77	£4,708.93	£4,850.19
NOI		£18,066.00	£18,607.98	£19,166.22	£19,741.21	£20,333.44	£20,943.45	£21,571.75	£22,218.90	£22,885.47	£23,572.03	£24,279.19
CapEx		£1,806.60	£1,860.80	£1,916.62	£1,974.12	£2,033.34	£2,094.34	£2,157.17	£2,221.89	£2,288.55	£2,357.20	
Sales price											£1,011,828.95	
Sales expenses											£50,591.45	
Property cash Flow	−£565,000.00	£16,259.40	£16,747.18	£17,249.60	£17,767.09	£18,300.10	£18,849.10	£19,414.57	£19,997.01	£20,596.92	£982,452.33	
Property IRR	8.06%											

Table 5.11 Income tax bands

2016 income tax rates (applied to earned income less personal allowance)		
Basic rate	20%	Up to £32,000
Higher rate	40%	£32,001 to £150,000
Additional rate	45%	Over £150,000

Source: Her Majesty's Revenue and Customs.

Table 5.12 Scheduled phase-out of the deductibility of financing charges

Tax Year[a]	Finance Costs Deductible from Rental Income	Basic Rate Tax Reduction
2016	100%	0%
2017	75%	25%
2018	50%	50%
2019	25%	75%
2020	0%	100%
2021 and beyond	0%	100%

[a] The tax year runs from April 6 to the following April 5. To simplify the calculation, this is being ignored.
Source: Her Majesty's Revenue and Customs.

Capital expenditures, like the purchase of the property or the cost of renovations beyond repairs for wear and tear, were not deductible.[10] The UK government was also implementing changes to income tax relief for residential property finance costs. Formerly, 100% of financing costs were deductible from rental income. This deduction was being phased out gradually beginning in 2017 (Table 5.12). In exchange, property investors would be allowed to reduce their overall tax liability with a basic rate tax reduction.

For many landlords, the change in the tax deductibility of interest charges would make no difference in the overall tax burden of owning investment property. For example, suppose a landlord without any other source of income generated £52,000 of rental income and incurred £20,000 in financing costs and £9000 of other allowable expenses. Under the old rules, taxable income would be calculated as £52,000 – £20,000 – £9000 = £23,000, which would incur a tax liability of £2400.[11] Once the new rules were completely implemented, taxable income would be calculated as £52,000 – £9000 = £43,000. This would incur a preliminary tax liability of £6400.[12] Then, the landlord would be able to apply a tax credit equal to 20% of the financing charge, which in this example is £4000. The total tax liability therefore would be £6400 – £4000 = £2400. Thus, landlords paying basic rates would not see an increase in their tax liability due to the new rules applied to financing charges.

However, property investors paying higher or additional rates would see their tax burden rise because they would lose an income deduction that would reduce higher or additional rates in exchange for a basic rate tax credit. For example, suppose a landlord with £35,000 of self-employment income earned an additional £18,000 in rental income, had £8000 in financing costs, and £2000 of allowable property expenses. Under the old rules, the landlord's total taxable income would be £35,000 + £18,000 – £8000 – £2000 = £43,000. As calculated previously, this would incur a tax liability of £6400. Once the new rules were implemented, taxable income would be £35,000 + £18,000 – £2000 = £51,000. This would incur a preliminary tax liability of £9600.[13] There would then be a tax credit applied equal to 20% × £8000 = £1600, for an overall tax bill of £8000. Thus, income taxes for this landlord would increase by £1600 following the implementation of the new rules.

118 *The taxation of property investment*

Stamp Duty Land Tax

A purchaser of residential property in the United Kingdom was subject to the residential stamp duty land tax (SDLT). Historically, this tax was assessed simply as 1% of the purchase price of the property in excess of £60,000.[14] With skyrocketing property prices in the heart of London, property buyers in 2 boroughs alone (Westminster and Kensington and Chelsea) paid nearly £1 billion in SDLT in the 2014 tax year.[15] In total, purchases of residential property contributed 43% of total SDLT UK revenue.[16] In December 2014, Her Majesty's Treasury replaced this constant SDLT tax rate with progressive brackets, which raised the marginal SDLT rate according to the value of the property (Table 5.13). In addition, a further 3% SDLT surcharge would apply to any residential purchase made by someone who already owned a residential property if that purchase occurred after November 26, 2015. Given Wilson's home in Kingston upon Thames, she would be subject to the additional SDLT surcharge.

Capital gains taxes

Finally, Wilson knew that a successful property investment would subject her to capital gains taxes – or so she hoped! Capital gains were calculated as the difference between what the owner sells the property for and what the owner originally paid for it. Taxpayers could deduct expenses associated with the buying or selling of the property (such as estate agents' and solicitors' fees) as well as the costs of improvement works. Normal operating expenses were not deductible for capital gains purposes. Higher- or additional-rate taxpayers would be subject to a 28% tax rate on all capital gains above an annual exempt amount (AEA) of £11,100. Basic rate taxpayers would face an 18% tax rate on gains that kept the taxpayer within the income bracket of basic rates. Additional gains would then be taxed at 28%.

Financing choices

Given the tax laws and the ways they were changing, Wilson realized that the fraction of her property's income owed to HMRC would depend on (1) the purchase and sale price of the investment, (2) the cost of any improvements made to the property, (3) the income the property generated over her investment horizon, and (4) the costs of financing the investment. She believed she would negotiate a fair price for the property, and she had already estimated likely improvements and the property's future sales price when analyzing the returns she hoped the property would deliver. Wilson knew that higher income meant higher taxes, but clearly higher income was still to be preferred. The remaining factor that would ultimately

Table 5.13 Stamp duty land tax

Property or Lease Premium or Transfer Value	SDLT Rate
Up to £125,000	0%
The next £125,000 (the portion from £125,001 to £250,000)	2%
The next £675,000 (the portion from £250,001 to £925,000)	5%
The next £575,000 (the portion from £925,001 to £1,500,000)	10%
The remaining amount (the portion above £1,500,000)	12%
. . . plus. . .	
an additional 3% if the property purchaser already owns a residential property and purchases another after November 26, 2015	

influence her tax liability was her financing costs. These costs were something over which she could exert at least some control.

Wilson was willing to commit £300,000 of savings to the property investment, although she wondered whether it would be a good idea to retain some of those savings in case of some unforeseen financial emergency. For the rest of the purchase price, she would have to borrow. Buy-to-let mortgages were readily available, but they came in many different varieties. Wilson didn't like the idea of getting a tracker mortgage, in which the interest rate was subject to change each month. Instead, she decided to focus on fixed-rate mortgages – in particular, mortgages whose interest rate was fixed for as long as possible, which in the current market meant 5 years. She outlined her financing choices under consideration (Table 5.14). Because she anticipated holding the property for longer than 5 years, Wilson could remortgage the flat at the prevailing buy-to-let rates at that time. Alternatively, she could keep whatever loan she chose today, understanding that regardless of her choice, her loan's interest rate would reset after 5 years to a rate equal to the sum of the Bank of England Base Rate (BEBR) and a margin of 4.49%. If Wilson's loan were resetting today, the reset rate would be 4.74%. Of course, 5 years from now the BEBR was likely to be higher than the current 0.25%, meaning her interest rate would reset to a rate higher than 4.74% if she did not remortgage (Table 5.15).

Table 5.14 Financing choices

	Repayment Mortgages		Interest-Only Mortgages	
LTV	60%	75%	60%	75%
Rate	2.37%	2.77%	2.47%	2.87%
Fixed period	5 years	5 years	5 years	5 years
Amortization	25	25	Interest only	Interest only
Reset rate	4.74%	4.74%	4.74%	4.74%

Table 5.15 Bank of England Base Rate

Year	Day	Month	Rate
2006	3	Aug	4.7500
	9	Nov	5.0000
2007	11	Jan	5.2500
	10	May	5.5000
	5	July	5.7500
	6	Dec	5.5000
2008	7	Feb	5.2500
	10	April	5.0000
	8	Oct	4.5000
	6	Nov	3.0000
	4	Dec	2.0000
2009	8	Jan	1.5000
	5	Feb	1.0000
	5	Mar	0.5000
2016	4	Aug	0.2500

Source: Bank of England, "Interest Rates and Bank Rate," accessed February 15, 2019, www.bankofengland.co.uk/monetary-policy/the-interest-rate-bank-rate.

120 *The taxation of property investment*

Next steps

With a tenant in place and a property management firm under contract, Wilson had only to negotiate a sales price and choose which financing offer to accept before officially becoming a property investor. Despite the recent changes to the tax rules for investors, she looked forward to a lifetime of wealth accumulation through property.

Notes

1. Real estate income is considered passive, and passive losses can be used to reduce passive gains. Wage income, however, is active, and so taxable wage income cannot be reduced by passive losses in a real estate investment portfolio.
2. Of course, the investor could undertake an additional exchange upon selling the retail center. To simplify the exposition, we assume that the investor pays all capital gains upon the sale of the second investment.
3. Note that the tax-related cash flows are independent of the performance of the new property. As such, they are reasonably certain. Therefore, it could be argued that the implicit borrowing rates should be compared to risk-free rates for a maturity equal to the holding period rather than the investor's borrowing rate which, as described in Chapter 3, depends among other things, on the borrower, the new property, and the borrower's financing choice.
4. The gross rental yield is calculated as the weekly rent × 52, divided by the purchase price.
5. The net rental yield is calculated as annual rental income less operating expenses, divided by the purchase price.
6. On average, Wilson had seen salary increases of 3% per year, which she expected to continue.
7. The size of the personal allowance tended to increase each year, although the magnitude and timing of such adjustments were difficult to predict.
8. Utilities and council tax (property tax) would be paid by Wilson's tenant.
9. Wilson had identified a property management firm that would cover all of the fees associated with leasing the property for a charge of 8% of total rental revenue.
10. In general, depreciation of investment property is not allowed in the United Kingdom. An exception would be the "wear and tear allowance" for furnished lettings. See "Renting Out Your Property (England and Wales)," www.gov.uk/renting-out-a-property (accessed February 15, 2019).
11. Calculated as (£23,000 – £11,000) × 20%. See Table 5.11.
12. Calculated as (£43,000 – £11,000) × 20%.
13. Calculated as (£32,000 × 20%) + (£8000 × 40%), since income above £43,000 is taxed at higher rates.
14. See Vanessa Houlder, "London's Status as Treasury Cash Cow Is Under Threat," *Financial Times*, September 6, 2016, www.ft.com/content/3360a24a-1b90-11e6-b286-cddde55ca122.
15. Ibid.
16. Ibid.

6 The quantification of risk

Introduction

In our discussion of property valuation in Chapter 2, we addressed the primary ways in which the risk of a property investment is incorporated into an investment decision. First, the development of our pro forma made adjustments to the cash flows so that the numbers in the spreadsheet were *expected* cash flows. That is, we explicitly recognized that leases were promises, and that it was necessary to adjust these promises for credit and vacancy losses, for example, to incorporate the risk that our tenants will not pay us every dollar they promised us on time. The second way in which our valuation incorporated risk was through the selection of the appropriate discount rate. Higher (systematic) risk warrants a higher discount rate. The selection of an appropriate discount rate allows for the calculation of the net present value (NPV) of an investment, and the NPV decision rule says that an investment that generates a positive NPV is a good one.

Having made the investment decision using the NPV decision rule, what still remains? The discount rate measures the expected return required for the investment, and as such is a measure of what you expect for the property's return *on average* to occur when you acquire a property at a price equal to the present value of its future cash flows. As an investor, you may still wish to know about outcomes that differ from the average outcome or about outcomes specific to your equity investment. It is natural to ponder questions such as: How will my investment perform if rental growth in the market isn't as strong as I expect? Will I lose money if it takes me longer than expected to fill my property's space when a tenant leaves? How likely is it that my realized return is 15% or higher?

All of these questions reflect risk in ways that cannot be measured by a single discount rate or by a single NPV. In this chapter, we describe additional methods by which property investors attempt to answer more general questions about the riskiness of their investment decisions. These various methods have specific purposes and have certain strengths and weaknesses of which all property investors should be aware.

Sensitivity to key assumptions

Continuing with our example of a 2-tenant industrial building that we saw in Chapter 5, the expected after-tax return to our equity investor was 12.26% per year over a 5-year investment horizon. Under our assumption that our pro forma was unbiased, the 12.26% return can be viewed as the return that our expected cash flows will generate.

Note that our pro forma being unbiased does not imply that the numbers we write down in a spreadsheet will be precisely equal to what we will realize in the future. Thus, an investor might reasonably wonder how the investment return will change if the realization of future

122 The quantification of risk

cash flows differs from what was written in the pro forma. For instance, in our example industrial property, we have assumed that rent in Space 1 will rise to $360,000 per year when that space is renewed in Year 3. What if, however, rent growth in the market turns out to be less strong than that and instead, when Year 3 comes around, you can only raise the rent to $330,000?

Of course, you can go into your spreadsheet and change the $360,000 per year being collected in Space 1 in Years 3 through 6 to $330,000 and see what happens to the after-tax return of the equity investor. If one does this, you will see that the equity after-tax return falls to 9.78% per year – a substantial drop. A **data table** is a display of "what if" calculations that describes how 1 variable – in this case, equity after-tax returns – is affected by changes in 1 or 2 other variables – in this case, rent growth for Space 1. Table 6.1 illustrates a 1-way data table, which means that only 1 variable in the spreadsheet model is changing. The bold row highlights what has been our benchmark assumption, that the rent collected in Space 1 will rise to $360,000 in Year 3.

From Table 6.1, you can see that the higher rent collected from Space 1 translates into higher returns for the equity investor. This is an intuitive relationship, but the data table quantifies the relationship so that the investor understands that each additional $30,000 in annual rent during Years 3 through 6 translates roughly into approximately a little more than a 2% higher annual return to the equity investor.

Data tables also allow an investor to consider changes to 2 parameters simultaneously. Continuing our example, suppose the investor also wanted to consider how after-tax returns would be affected by exit cap rates different from the 8% assumed in the model thus far. A 2-way data table can be constructed that calculates the equity after-tax return the investor would receive for each pairwise realization of Space 1 future rent and the property's exit cap rate. These calculations are shown in Table 6.2.

Notice in Table 6.2, the center column is simply a repeat of Table 6.1 because it holds the exit cap rate constant at 8%, which is our originally assumed value. However, moving across

Table 6.1 Impact of rent growth on after-tax equity returns

Rent in Space 1 beginning in Year 3	Equity After-Tax IRR
$300,000.00	7.06%
$330,000.00	9.78%
$360,000.00	**12.26%**
$390,000.00	14.54%
$420,000.00	16.66%

Table 6.2 Impact of rent growth and exit cap on after-tax equity returns

		Exit cap rate				
		6.00%	7.00%	**8.00%**	9.00%	10.00%
	$300,000	16.88%	11.69%	7.06%	2.82%	–1.15%
	$330,000	19.40%	14.30%	9.78%	5.67%	1.87%
Rent in Space 1 beginning in Year 3	**$360,000**	21.73%	16.69%	**12.26%**	8.26%	4.58%
	$390,000	23.90%	18.91%	14.54%	10.62%	7.04%
	$420,000	25.92%	20.98%	16.66%	12.81%	9.30%

columns demonstrates the influence that exit cap rates have on realized returns. Holding the rent in Space 1 constant at $360,000, an increase in the exit cap rate from 8% to 9% reduces the equity after-tax return from 12.26% to 8.26%.

As is evident from Tables 6.1 and 6.2, data tables can be a useful way to explore how 1 or 2 underlying assumptions influence some ultimate variable of interest, which in this example is the annual return received by the equity investor after tax. The greatest advantage of using data tables to quantify risk is that constructing such tables is easy to do. In addition, data tables offer the advantage that they are easy to explain and interpret.

Despite these important advantages, there are some noteworthy disadvantages to using data tables to communicate risks in a property investment. The most obvious is that you are only able to analyze at most 2 variables at the same time. For example, suppose as an investor you were worried not only about the rent in Space 1 and the exit cap rate but also about the interest rate that you would have to pay on your mortgage debt. Of course, you could construct multiple 2-way data tables for each possible interest rate, but before long the number of tables becomes unmanageable.

A second and more significant disadvantage to using data tables to quantify risk is that the data reported is often influenced by the biases of the person constructing the table. Consider again Table 6.2. If you look closely at the numbers, it seems to suggest that unexpected movements in exit cap rates are more relevant to ultimate returns than movements in Space 1 rent. While this may, in fact, be true, consider the alternative data table constructed for the same 2 variables shown in Table 6.3.

The only differences between Table 6.2 and Table 6.3 are the values that Space 1 rent and the exit cap rate are allowed to have. Otherwise, the calculations are identical. However, if you look at Table 6.3, it appears that Space 1 rent is a more significant determinant of equity returns than exit cap rates. This is simply an artifact of the choice to limit the range of exit cap values, while at the same time increase the range of rental rates. Nothing has fundamentally changed about the relationship between rents, exit caps, and returns. All that has changed is how information about that relationship is displayed.

This simple example illustrates that what someone takes from a data table is influenced by subjective choices made by the person who constructs the table. Why might this be a problem? Suppose that you are being asked by a general partner of a real estate deal to invest in this particular property. The general partner claims that your return is not particularly sensitive to the exit cap rate assumption and shows you Table 6.3 as "proof." This data table does seem to suggest that returns are not very strongly affected by any cap rate *in the table*. As an investor, you have to be cognizant of the fact that the table was constructed with a particular purpose in mind, namely to make the investor not particularly worried about the exit cap rate assumption.

Table 6.3 A different view of the impact of rent growth and exit cap

		Exit cap rate				
		7.750%	7.875%	**8.000%**	8.125%	8.250%
	$260,000	4.13%	3.54%	2.95%	2.37%	1.80%
	$310,000	9.10%	8.54%	8.00%	7.45%	6.92%
Rent in Space 1 beginning in Year 3	**$360,000**	13.32%	12.79%	**12.26%**	11.74%	11.22%
	$410,000	17.01%	16.49%	15.97%	15.46%	14.97%
	$460,000	20.29%	19.77%	19.27%	18.77%	18.28%

124 *The quantification of risk*

Table 6.4 Pros and cons of using data tables

Advantages	Disadvantages
• Easy and convenient method for displaying output sensitivity • Used across many industries • Target audience understands method	• Not probability weighted • Limited to analysis of the effects of changing a maximum of 2 variables at once • Vulnerable to analyst's cognitive biases

This preceding discussion suggests the last major disadvantage of using data tables as a means to quantify risk in a property investment. Data tables do not convey information regarding the relative likelihood of any particular outcome illustrated in the table. Although it is relatively common to construct data tables as we have done, with the baseline assumption in the center, it is not necessary (and in fact, unlikely) that the adjacent values are equally likely. This is because the variables of interest do not move independently of one another. For example, rising cap rates tend to indicate a softening of real estate markets that might be generally associated with lower rent growth. Lower cap rates, by contrast, would typically accompany higher rent growth. Therefore, it is likely that values in the upper right and lower left of the data table are more likely to occur than values in the upper left (low cap rates and low rent growth) and lower right (high cap rates and high rent growth). For this reason, one simply cannot count the number or fraction of agreeable returns in a data table to learn something about the likelihood that agreeable returns will be forthcoming.

A summary of the pros and cons of using data tables is provided in Table 6.4.

Scenario analysis

Our previous discussion regarding data tables highlighted 3 key problems with data tables as a way to quantify risk in a real estate investment. First, you are limited to only 2 variables to explore at a time. Second, you are subject to the potential biases of the person constructing the table. Third, the tables lack any connection to probabilities of the various outcomes. In this section, we describe how scenario analysis can address the first 2 deficiencies of data tables. The following section discusses how Monte Carlo analysis can address the third deficiency.

Scenario analysis allows for the calculation of a relevant metric, say equity after-tax return, after any number of input parameters are changed. This is particularly useful when you have in mind a more complicated outcome that cannot be summarized in 1 or 2 assumptions. For instance, in the last section, we contemplated how investor returns were affected by simultaneous changes to exit cap rates and the future rent collected in Space 1. We also considered how we might worry about the interest rate on our mortgage. Rather than construct a series of data tables for each possible level of a mortgage rate, a scenario analysis allows us to consider any number of possible changes to our benchmark pro forma analysis simultaneously. In fact, any number of variables can be changed within a scenario.

For example, consider the 3 scenarios outlined in Table 6.5. In most scenario analysis, one particular outcome is typically highlighted. In our pro forma analysis, we have highlighted cash flows based on expected future values. It is these expected future cash flows that are the necessary inputs into a discounted cash flow (DCF) valuation of the property. For this reason, I will illustrate scenario analysis beginning from these same assumptions. Thus, my benchmark scenario is defined as an outcome where rental growth in Space 1, the property exit cap rate, the mortgage rate, and the TI expense are set equal to their expected

Table 6.5 Scenario definitions

Assumptions:	Benchmark	Negative	Positive
Rental growth	$360,000.00	$330,000.00	$390,000.00
Exit cap	8.00%	10.0%	6.00%
Mortgage rate	4.25%	4.75%	3.75%
TIs	$60,000.00	$80,000.00	$40,000.00

Table 6.6 Data table of scenarios

Scenario	Equity after tax return
Benchmark	12.26%
Negative	1.05%
Positive	24.51%

values. I then define 2 additional scenarios, where each of these 4 variables move in a direction that would benefit the investor (the positive scenario) or would hurt the investor (the negative scenario). Note, too, that I have defined the scenarios to be symmetric around the benchmark.

Once these scenarios are defined, then one can explore how the variable of interest, for example, the investor's after-tax rate of return, varies across scenarios. Thus, unlike data tables, scenario analysis can account for changes in any number of inputs simultaneously. With the scenario assumptions given in Table 6.5, one can use a data table across scenarios as shown in Table 6.6 to calculate how investor returns would vary across any number of scenarios.

Like data tables, scenario analysis is easy to explain. However, as illustrated, scenario analyses provide useful advantages over data tables for investors wishing to quantify the risk they face when making a real estate investment. First, scenarios can be constructed that involve changes to any number of assumptions. Second, scenario analyses can alleviate concerns about the biases implicit in data table construction because a scenario can be customized to *any* investor's concerns. For instance, in an equity limited partnership, a GP might present its potential partners with a benchmark return and potentially a data table or two. In response, an LP can ask for any specific scenario to be considered. If an investor is worried about a particular confluence of events that leads to inputs behaving in a particular way, a scenario with these exact features can be created and its overall impact determined. Thus, scenario analysis can be used by outside investors to illustrate a wider range of possible outcomes than might have otherwise been considered by inside investors.

Despite these advantages of scenario analysis, there remain a number of disadvantages to using scenario analysis as a method to quantify risk in a real estate investment. First, like the use of data tables, scenario analysis ignores the relative likelihood of any given outcome. Put another way, any given scenario is extremely unlikely to occur precisely given that there are an infinite number of scenarios that one *could* consider. As a result, one may falsely assume that all scenarios presented are equally likely to occur. Relatedly, the presentation of *particular* scenarios continues to reflect any biases of those who propose particular scenarios. That is, it may be easy for an investor to assume that the scenarios being considered reflect all the meaningful possibilities, when there always remains a wide set of alternative outcomes.

126 *The quantification of risk*

Table 6.7 Pros and cons of using scenario analysis

Advantages	Disadvantages
• Easy to understand and explain • If a specific scenario is of particular concern to management/investors it can be explicitly tested • Prescriptive in nature – each specific input value can be explained and examined with respect to its impact on the bottom line • Can measure the impact of many variables at the same time	• Not probability weighted – all scenarios implicitly equally likely • Ignores the effect of correlation across variables • Lacking robustness – if the variables were slightly different, you may get a totally different result • Problematic when assessing the effects of leverage – default risk may be masked by errors in scenario choice • Examining scenarios separately ignores the possibility that aspects of each scenario occur simultaneously

Second, the methodology is somewhat lacking in robustness in that looking at scenarios independently cannot tell you what will happen if aspects of different scenarios were to occur simultaneously. Similarly, scenario analysis fails to tell you what might happen if the actual outcome is slightly different from what was assumed in the scenario.

A summary of the pros and cons of using scenario analysis is provided in Table 6.7.

Monte Carlo analysis

Our previous discussion of data tables and scenario analysis was useful in answering questions of the "What if?" variety. That is, these tools to quantify risk are most useful if you want to know how investor returns would vary as input assumptions change. However, these tools are not particularly useful in asking questions of the "How likely?" variety. That is, an investor might reasonably want to answer questions such as: How likely is it that I lose money on this investment? How likely is it that my returns exceed 15%? How likely is it that the general partner (GP) will earn a promote? Monte Carlo analysis provides answers to these types of questions.

Monte Carlo analysis works quite differently than either data tables or scenario analysis. In the previous sections, we saw that sensitivity analysis and scenarios can help measure an output *given* particular values of the inputs. Monte Carlo works rather differently. Rather than assume any given set of input values, a Monte Carlo analysis assumes that inputs are drawn from probability distributions and, for each draw, outputs are measured. By drawing multiple (thousands of) times from these distributions, a distribution of outcomes is calculated. Therefore, the output of Monte Carlo analysis is not a single value, but rather a probability distribution that allows the answering of "How likely?" questions.

To illustrate how Monte Carlo analysis can be used to quantify risk, consider the 4 variables of particular interest to our analysis of risk for the industrial property: the rental growth in Space 1, the exit cap rate for the property, the mortgage interest rate, and the level of the TI expense. Our assumptions in our scenario analysis were outlined in Table 6.5. In a Monte Carlo analysis, we do not specify precise values for our variables of interest, but rather we assume that these variables are chosen from specified distributions. Then, we make repeated draws from our chosen distributions and calculate our outcome variable (e.g. after-tax equity rate of return) for each draw.

Suppose, for example, that we can quantify our uncertainty about the rent we will collect from Space 1 in Year 3 by a triangular probability distribution. This distribution was chosen to have a minimum, peak, and maximum value of $300,000, $350,000, and $430,000, respectively. With these parameters, the expected rental rate will be $360,000, which matches the assumption in our benchmark scenario. We can then choose distributions for our other 3 variables. For instance, we might assume that exit cap rates are equal to 5.5% plus a lognormally distributed variable, with a mean of 2.5% and a standard deviation of 4%. This assumption again implies a mean exit cap rate of 8%, matching our benchmark assumptions. It also implies that the exit cap rate on the property will never fall below 5.5% but might potentially (with low probability) be extremely high. The potential for extremely high exit cap rates is meant to replicate the real-world possibility that investment property prices might substantially fall during a severe market downturn. Mortgage interest rates typically move in discrete amounts, so we can assume that with probability 0.6 the interest rate will be our benchmark of 4.25%, but with probability 0.2 the interest rate might be either 4.00% or 4.50% instead. In this example, I do not allow much uncertainty in the interest rate because interest rates are typically locked at or near the time of property acquisition. Over such short time horizons, mortgages rates are unlikely to move much. Finally, we may assume that the TI expense we occur in both Year 1 and Year 3 is distributed lognormally, yet still has an expected value of $60,000. The parameters of the lognormal distribution were chosen so that TI expenses would never be less than $40,000 and would only exceed $100,000 with 5% probability. These distribution assumptions are summarized in Table 6.8 and visualized in Figure 6.1. Here, the choice of lognormal distribution is meant to represent that most of the time, TI expenses are between $40,000 and $80,000, but in some cases might need to be significantly higher to attract and meet the specific needs of a specific tenant.

To see how a Monte Carlo analysis works, we simply draw a value from each of these 4 distributions and calculate the after-tax return to our equity investor. Then we repeat this numerous times. For example, after completing a 10,000-draw analysis, the distribution of equity after-tax returns is shown in Figure 6.2.

Figure 6.2 demonstrates the benefits of a Monte Carlo analysis relative to our previous approaches to quantifying risk. The output from such an analysis allows investors to quantify risk in a probabilistic sense. For example, the simulations generate a mean return of 13.2%, a bit higher than our findings in the benchmark scenario. The maximum return across the 10,000 draws was 28.7%, and slightly more than 1% of the time our investor would have lost the entire amount invested. With a distribution of results, the Monte Carlo simulations allow for any of the "How likely?" questions that investors may have. For instance, only 5% of the time did the equity investor earn less than -8.7%, and 5% of the time the equity investor earned in excess of 24.5%.

Table 6.8 Monte Carlo distribution assumptions

Parameter	Distribution Assumptions
Rent in Space 1 beginning in Year 3	Triangular with Min $300,000, Max $430,000, Mode $350,000, Mean $360,000
Exit cap rate	Lognormal, with Mean 2.5%, Standard Deviation of 4%, with a shift of 5.5%
Mortgage rate	Discrete: 4.0% with probability 0.2, 4.25% with probability 0.6, 4.5% with probability 0.2
TIs	Lognormal, with Mean $20,000, Standard Deviation $35,000, with a shift of $40,000

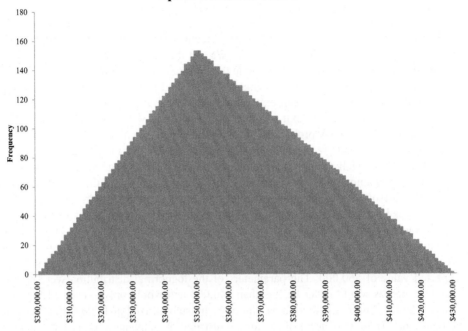
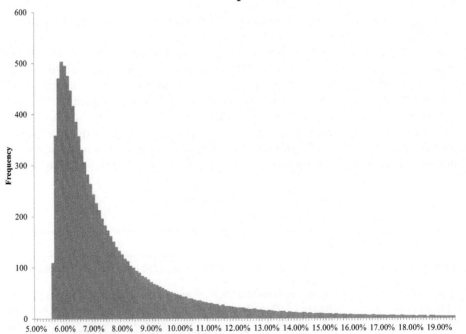

Figure 6.1 A visualization of the input parameters

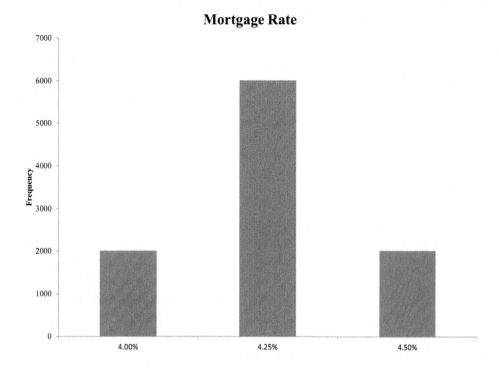
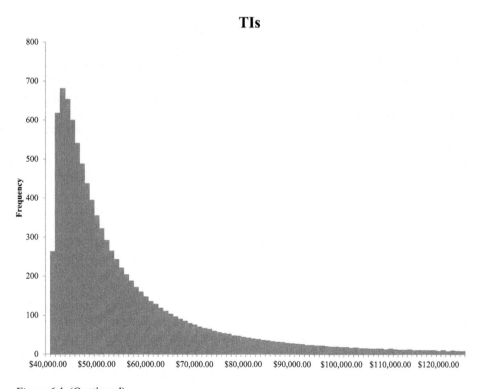

Figure 6.1 (Continued)

130 *The quantification of risk*

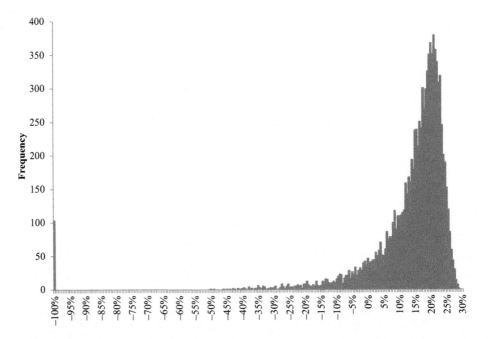

Figure 6.2 Equity after-tax return from 10,000 simulations

Monte Carlo analysis also allows measurement of the most significant sources of variation in the outcome variable. In our example, one can measure which of the 4 input variables was the most significant determinant of equity after-tax returns. For example, suppose we sort our simulation output by the exit cap rate that was chosen in each draw. Then consider grouping our output variable, the equity after-tax return, into 10 groups, where each group are deciles created from the data sorted by the exit cap. Finally, calculate the average after after-tax return in each decile. In my simulation, the average after-tax return for simulations where the exit cap was in its lowest decile was 23.3%. Conversely, the average after-tax return for simulations where the exit cap was in its highest decile was -19.4%. Analogous calculations can be conducted by repeating this procedure after sorting the simulation data by each of the other 3 inputs. Then, the average value of the output variable can be displayed for the highest and lowest decile for each variable. The result of this procedure is depicted in Figure 6.3. This "tornado" graphic gives the investor important insight into the input variable that most influences the output variable of interest. In this stylized example, it is not surprising that the exit cap rate is by far the most significant variable in determining equity after-tax returns.

Another key advantage of Monte Carlo analysis is that it can incorporate the reality that economic conditions typically drive many variables in predictable ways. That is, rent growth and cap rates are both impacted by economic conditions. If conditions are good, then rent growth tends to increase and cap rates tend to fall. Thus, rent growth and cap rates are negatively correlated. In the Monte Carlo analysis completed previously, it was assumed that all 4 variables were independent of one another, which is likely not the case. One can change this assumption and build in the expectation that the input variables move together as expected. To illustrate the significance of correlation assumptions, we can repeat the analysis

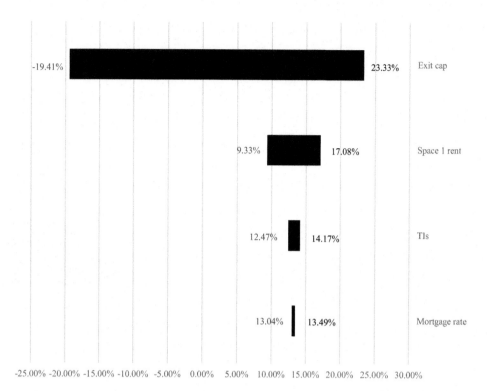

Figure 6.3 Measuring the influence of input variables

Table 6.9 Assumed correlations between variables

	Rent in Space 1 Beginning in Year 3	Exit Cap Rate	Mortgage Rate	TIs
Rent in Space 1 beginning in Year 3	1			
Exit cap rate	−0.3	1		
Mortgage rate	0.2	−0.1	1	
TIs	0	0	0	1

conducted above while introducing additional assumptions regarding how the input variables move together. Another 10,000 draws were made from the same distributions, only now requiring that the variables have the correlations described in Table 6.9. Note that we have continued to maintain the assumption that TI expenses are independent of the other inputs.

When the simulation includes correlation between the input variables, then the equity after-tax returns tend to be more widely dispersed. 90% of the simulated after-tax returns are between -10.1% and 25.2% as compared to -8.7% and 24.5%, mentioned previously. This is a direct result of the fact that we have assumed that the input variables move together. That is, since higher rent growth accompanies lower cap rates, returns tend to be boosted in good draws, whereas lower rent growth accompanies higher cap rates, which tends to lower returns in bad draws. Figure 6.4 summarizes the simulation output

132   The quantification of risk

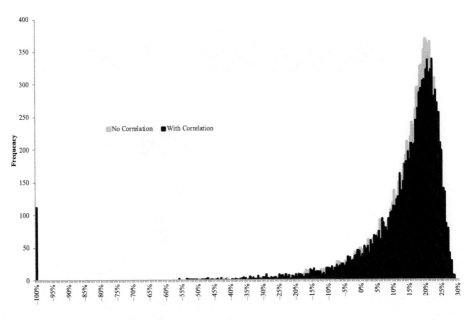

Figure 6.4 The impact of correlation on equity returns

with and without correlation assumptions. Thus, embedding correlations between input variables can influence how one measures the likelihood of achieving either very high or very low returns.

This discussion has illustrated the potential advantages of using Monte Carlo analysis to quantify risk of a real estate investment. In particular, investors using such an analysis can answer questions about the likelihood of certain outcomes happening. This is fundamentally different from scenarios and data tables, which were prescriptive in that they only told you what would happen for any given set of assumptions. Instead, Monte Carlo analysis calculates a probability distribution that represents the combined impacts of all aspects of the financial model. Thus, the results of such analysis are more robust that those of scenario analysis because outcomes can be weighted by their probabilities. As computer software has made running simulations relatively simple, conducting Monte Carlo analysis may soon be as routine as data tables are today.

On the downside, meaningful implementation of Monte Carlo would require selecting appropriate distributions for all the relevant input variables. In practice, this means making a distributional assumption that fits the data. Similarly, identifying relevant correlations and estimating them with data would be crucial. The output of a simulation can only be as good as the quality of the assumptions of the model. Here too, however, it is likely that investors' ability to make quality assumptions will increase when better real estate data is available. Finally, the benefit of probabilistic outputs in a Monte Carlo simulation is partially offset by the method's inability to explain why certain outcomes occur. That is, this approach to quantifying risk can identify the probability of an outcome, but not the chain of economic events that led to the outcome being realized.

A summary of the pros and cons of using scenario analysis is provided in Table 6.10.

Table 6.10 Pros and cons of using Monte Carlo analysis

Advantages	Disadvantages
• Avoids the difficulty of specifying input values • Generates more robust results than scenario analysis, and weights results according to probability of occurrence • Calculates a probability distribution that represents the combined impacts of the model's total uncertainty • Allows for impact of correlation across variables • Computer software has made simulation easier and more accessible	• Long time horizon for real estate investment leads to high variance of results across the forecast period • Model is only as good as the historical distributions that can be calculated for input variables • Lack of deterministic description with the results – method can identify the probability of an outcome but not the chain of events that led to the outcome

Now what?

This chapter has outlined a number of different approaches that real estate investors can use to quantify the risk they face. Having completed such analysis, what then? In light of the risks, investors can choose to *avoid* the risk by not making the investment. They might *transfer* some or part of the risk by buying various forms of insurance. They might *mitigate* some of the risk by either using hedging instruments available in financial markets or by making certain operational decisions that tend to reduce risk. Finally, an investor knowledgeable about the risks being faced can choose to *accept* those risks and continue to invest as planned.

Note, too, that a thorough risk assessment of a real estate investment is likely more involved than the stylized example that was considered in this chapter. That is because there are a number of risks that real estate investors face that cannot be easily quantified in an input variable in a spreadsheet. In this last section of the chapter, we outline some of these risks, along with how these risks might be addressed.

Liquidity risk

Real estate is an illiquid asset. This means that any given piece of property trades only very rarely. Further, the frequency of trade is going to vary according to the strength of the local market, the desirability of the particular asset in its market, and the asset's physical condition. Similarly, each individual property has unique features that make it a challenge to understand its true value.

Investors who place great concern over a property's liquidity tend to invest in higher quality, "Grade A" assets. Similarly, they tend to invest in more active markets. By doing so, they often sacrifice expected returns in exchange for higher liquidity. Investors worried about liquidity may also use less debt to minimize the likelihood of being forced to sell during a downturn.

Tenant risk

Obviously, one's analysis of a real estate investment must reflect the likelihood of a tenant's ability to pay rent. Tenants failing to fulfill rent obligations cause investors to incur costs to take legal actions or to evict and re-tenant the property.

Investors particularly concerned about tenant risk ideally lease to higher quality tenants. Of course, higher quality tenants might be expected to negotiate better lease terms and be willing to only lease space in top quality markets and properties. Obviously, mitigating tenant risk involves investors' completing due diligence on prospective tenants. At one extreme, tenants might be strong corporations with a lot of publicly available information on their creditworthiness. At the other, a tenant may have poor or no credit history whatsoever.

Sector and geographic risk

Different real estate sectors, like office, retail, and housing, present different risks and performance characteristics. Hotels, for example, have overnight leases and are very cyclical. Office buildings, by contrast, typically have longer leases, and as such typically provide more stable cash flow. One way to diversify sector risk would be to invest in properties in multiple sectors, yet in doing so one makes it more difficult to gain the managerial expertise that might be sector specific.

Likewise, different locations present different risk profiles that are difficult to mitigate in ways other than through the decision of where to invest. A real estate portfolio that is more geographically diversified mitigates location risk, but like sector risk makes it difficult to take advantage of particular local knowledge.

Supply risk

Certain markets have geographic features or regulatory planning and zoning laws that affect the construction of property that might compete with your proposed investment. It is difficult to address supply risk other than through your selection of where to invest.

Risk assessment in real estate: should I rent my condo?

In this case, you are put in the shoes of Diana Mulhall, a risk management professional in the midst of a career move that eliminates her need to own a condo in downtown Toronto. With the Canadian city's real estate market being hot, Mulhall wonders whether she should sell the condo or keep it as a rental property. In completing this case, you will be able to:

1 Learn about how to measure risk in a real estate context in ways beyond estimating a risk-adjusted discount rate.
2 Construct a probabilistic analysis of real estate returns by incorporating Monte Carlo simulation into a real estate pro forma.
3 Compare and contrast the relative strengths and weaknesses of probabilistic and deterministic risk analyses.
4 Understand the effects of foreign currency transaction exposure on expected returns denominated in domestic currency and qualitatively discuss the merits of hedging foreign currency risk.

Should I rent my condo: assessing risks of a property investment

In early 2018, Diana Mulhall faced a financial decision that she hadn't anticipated. She had just decided to leave the consulting firm where she had spent the better part of her career to start her own risk assessment advisory business in her hometown of Chicago. For the past 12 years, she had climbed the ranks at management consultancy Accent, finally becoming a partner in the firm's risk management division 3 years ago. Her major client was the Canadian Imperial Bank of Commerce (CIBC). She eventually grew tired of spending 4 nights a week in a Toronto hotel, so in 2013 she purchased a small, 1-bedroom condo in Toronto's financial district that was within walking distance of the bank. Being able to return in the evenings to a condo decorated to her own tastes improved her mood during her stays in Toronto. Now that her career move would put her regular Toronto commute to an end, Mulhall's first instinct was to sell the condo, which had soared in value in Toronto's booming housing market. Perhaps, though, that was a hasty decision, and she should consider keeping the property as an investment, renting it out to capitalize on its prime location in a booming city. As part of her portfolio of assets, Mulhall already owned a number of investment properties, but all of those were located in Chicago. Perhaps the Toronto condo would add useful geographic diversification? It had certainly been a great investment up to this point. But what about the risks? Fortunately, risk was her business during all of her years at Accent. She knew all about the quantification of risk. It was time to apply her skills for her personal benefit.

The property

The Toronto condominium was located in the 35-story boutique condo building at 21 Nelson Street in downtown Toronto (Figure 6.5). It was only about a kilometer from CIBC headquarters in Commerce Court and around the corner from the Shangri-La Hotel, at which Mulhall had spent over 300 nights in her years commuting to the city. The building offered luxury, loft-style condo living, with 24-hour concierge service, exercise facilities, a rooftop bar, outdoor barbecue, and even its own yoga studio. Mulhall's junior 1-bedroom unit was located on the 11th floor and offered a view of the CN Tower. The living room, kitchen, and powder room were on the entry level, and then a staircase led to the sleeping area and full bathroom upstairs (Figure 6.6). In total, the unit was 606 square feet, and Mulhall had purchased the unit in July 2013 for $374,000.[1]

Figure 6.5 The boutique condos at 21 Nelson Street
Source: http://www.21nelson.com/building/

136 *The quantification of risk*

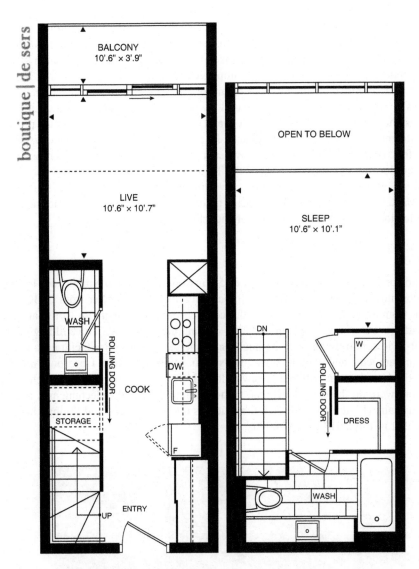

Figure 6.6 Floor plan of Diana Mulhall's condo
Source: http://www.21nelson.com/suites/

The neighborhood

The city of Toronto is divided into 140 neighborhoods. Mulhall's condominium was in neighborhood 77, the Waterfront Communities – The Island, so named due to its proximity to Lake Ontario. The population in the area had increased 52% between 2011 and 2016 (Figure 6.7), with 68% of its population being working age. The business district and waterfront location tended to attract wealthier households relative to Toronto overall (Figure 6.8). Being in the heart of downtown Toronto earned the area a 100 out of 100 transit score for ease of access to public transportation and a 99 out of 100 walk score, because nearly everything one might want was within walking distance.

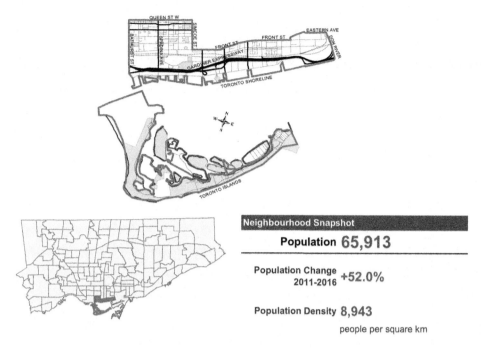

Figure 6.7 The Waterfront Communities – The Island neighborhood
Source: Statistics Canada, 2016 Census of Population.

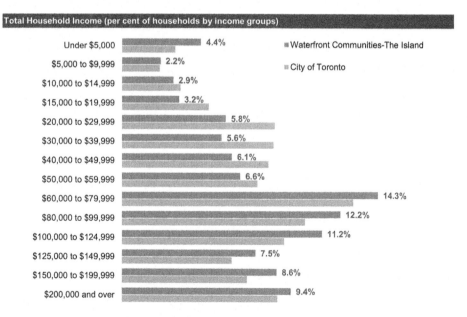

Figure 6.8 Income distribution of the neighborhood
Source: Statistics Canada, 2016 Census of Population.

138 The quantification of risk

Market trends

Toronto real estate had been a fabulous investment over the past decade. Rents had been climbing since 2006 at an average rate of 2.7% per year (Figure 6.9). Vacancy was virtually nonexistent, as the demand for space in central Toronto seemed consistently to outstrip supply (Figure 6.10). Institutional investment in Toronto also indicated a strong market. Cap rates for multifamily properties had been falling for many years (Figure 6.11).

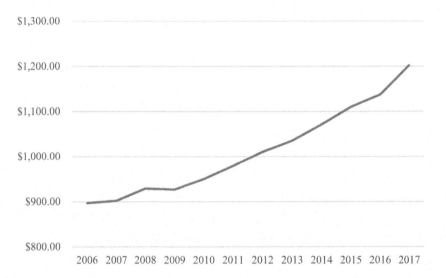

Figure 6.9 Average monthly rent for a 1-bedroom apartment in Toronto (Center City)
Source: Canada Mortgage and Housing Corporation.

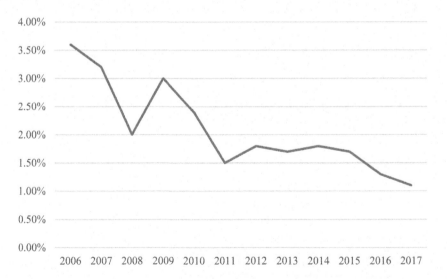

Figure 6.10 Vacancy rates for a 1-bedroom apartment in Toronto (Center City)
Source: Canada Mortgage and Housing Corporation.

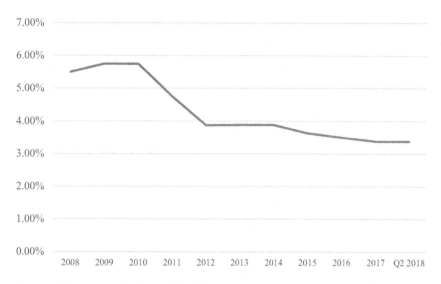

Figure 6.11 Cap rates for institutional investment in greater Toronto area, high-rise multifamily A
Source: CBRE.

Choices

Selling the condo would be the simplest choice for Mulhall. To get an idea of what the condo might be worth, she called a local leasing broker, who sent her all of the sales in her building for the past few years (Table 6.11). Rather than selling, however, it might benefit Mulhall's overall portfolio to maintain ownership of the condo and turn it into a rental property. The broker also provided her with all recent leases signed at the property (Table 6.12) so that Mulhall could get a sense of the rental market. If she leased the unit, it meant that she would have a steady source of income from the property. Of course, it would also mean having a steady source of expenses. She took a few moments to jot down the expenses she was currently incurring on the property (Table 6.13). Mulhall assumed that if she leased, she would likely hold the condo for an additional 10 years. With some benchmark assumptions regarding the condo's average vacancy, ultimate sales price, rent growth, and the growth rate of expenses, it wasn't difficult for Mulhall to determine the rate of return her condo would deliver as an investment. Mulhall knew from her years of experience that any decent analyst can make an investment thesis look good in a spreadsheet. The key, she knew, was to be careful with her assumptions.

With these benchmark assumptions aside, Mulhall paused to consider the key risks of the investment. First, it was obvious to her that her return forecast was closely related to her assumed sales price in 10 years. Second, it was also clear that the property's rental rate growth would be influential. Third, her ability to keep the property leased would be important. Given her expertise in risk measurement and management, Mulhall was quite confident of her ability to apply sophisticated risk measurement tools to her specific property investment. She also paused to consider how important it was that the condo was in Toronto and her rental payments would be received in a foreign currency. Her years of commuting across the border gave her an awareness of currency fluctuations that would be helpful (Figure 6.12).

Table 6.11 Recent sales at 21 Nelson St.

Date	Unit	Square Feet	Price	Days on Market
Aug-18	1105	0–499	$475,000	13
Jun-18	302	700–799	$640,000	9
Jun-18	702	700–799	$667,000	20
Jun-18	622	0–499	$408,000	3
Jun-18	514	700–799	$637,000	6
May-18	LPH29	600–699	$595,000	8
May-18	221	0–499	$375,000	5
May-18	UPH1	1000–1199	$986,500	6
May-18	523	700–799	$642,500	18
Apr-18	907	900–999	$870,000	5
Apr-18	413	700–799	$595,000	11
Mar-18	906	900–999	$830,000	1
Mar-18	818	1000–1199	$870,000	5
Dec-17	1022	700–799	$618,000	3
Dec-17	223	800–899	$590,000	21
Dec-17	216	700–799	$585,000	6
Dec-17	508	600–699	$468,000	7
Dec-17	212	700–799	$581,800	5
Nov-17	402	700–799	$622,500	4
Nov-17	315	600–699	$515,000	13
Nov-17	1106	800–899	$720,000	7
Oct-17	LPH13	1000–1199	$800,000	10
Oct-17	1018	1000–1199	$850,000	4
Sep-17	LPH15	600–699	$550,500	15
Sep-17	913	600–699	$538,000	14
Aug-17	923	600–699	$518,000	4
Aug-17	608	600–699	$450,000	3
Jun-17	323	700–799	$565,000	24
May-17	UPH10	1400–1599	$1,026,000	18
May-17	426	500–599	$386,000	5
May-17	624	0–499	$335,000	10
May-17	425	1000–1199	$720,000	28
Apr-17	1107	900–999	$715,000	8
Apr-17	LPH25	1000–1199	$800,000	4
Mar-17	327	700–799	$480,000	17
Feb-17	UPH19	1200–1399	$860,000	22
Jan-17	LPH20	1200–1399	$750,000	5
Dec-16	325	900–999	$615,000	19
Dec-16	230	500–599	$355,000	9
Oct-16	306	700–799	$407,000	40
Oct-16	627	700–799	$475,000	5
Sep-16	328	500–599	$375,000	100
Sep-16	1122	700–799	$510,000	4
Sep-16	326	500–599	$330,000	7
Sep-16	LPH28	600–699	$425,000	2
Sep-16	328	500–599	$375,000	13
Aug-16	LPH9	1000–1199	$625,000	1
Jul-16	615	600–699	$417,000	34
Jul-16	LPH12	1000–1199	$615,000	0
Jun-16	726	500–599	$340,000	5
May-16	218	0–499	$274,000	8
May-16	922	700–799	$485,000	16
Apr-16	318	0–499	$267,000	7

Date	Unit	Square Feet	Price	Days on Market
Apr-16	LPH7	900–999	$530,000	39
Mar-16	1118	1000–1199	$699,900	11
Mar-16	LPH4	1200–1399	$865,000	13
Mar-16	LPH32	1000–1199	$638,000	74
Mar-16	1119	500–599	$358,000	4
Mar-16	UPH8	800–899	$535,000	7
Mar-16	LPH25	1000–1199	$614,640	56
Jan-16	1020	700–799	$468,800	59
Jan-16	UPH16	1400–1599	$905,000	21
Dec-15	914	600–699	$390,000	27
Dec-15	UPH20	800–899	$574,463	138
Dec-15	1001	700–799	$425,000	28
Nov-15	729	700–799	$457,000	23
Nov-15	802	700–799	$475,000	23
Nov-15	UPH15	1600–1799	$960,000	14
Nov-15	204	0–499	$255,000	46
Oct-15	1121	700–799	$463,800	34
Oct-15	1103	700–799	$493,500	15
Oct-15	817	1000–1199	$760,000	29
Oct-15	307	500–599	$322,500	36
Sep-15	LPH29	600–699	$418,000	7
Sep-15	531	700–799	$418,000	64

Source: condos.ca.

Table 6.12 Recent leases at 21 Nelson St.

Date	Unit	Square Feet	Rent	Days on Market
Aug-18	1102	700–799	$3100	1
Aug-18	221	0–499	$2050	1
Aug-18	1111	500–599	$2100	0
Aug-18	622	0–499	$2250	20
Aug-18	610	500–599	$2050	1
Jul-18	LPH2	1200–1399	$4400	40
Jul-18	322	0–499	$2100	0
Jul-18	226	500–599	$2175	0
Jun-18	718	0–499	$1870	2
Jun-18	LPH12	1000–1199	$3500	1
Jun-18	203	700–799	$3000	6
Jun-18	321	0–499	$1950	26
Jun-18	LPH28	600–699	$2450	0
May-18	915	600–699	$2450	0
May-18	1011	500–599	$2100	0
May-18	517	0–499	$1850	14
May-18	316	700–799	$2700	1
Apr-18	LPH1	1200–1399	$4300	73
Apr-18	UPH5	1600–1799	$5750	28
Apr-18	821	700–799	$2640	1
Apr-18	723	700–799	$2400	0
Apr-18	905	0–499	$2100	0
Mar-18	314	700–799	$2650	0

(*Continued*)

Table 6.12 (Continued)

Date	Unit	Square Feet	Rent	Days on Market
Feb-18	627	700–799	$2450	0
Feb-18	414	700–799	$2550	0
Feb-18	216	700–799	$2600	3
Jan-18	403	700–799	$2700	25
Jan-18	804	700–799	$2595	13
Jan-18	1003	700–799	$2700	0
Jan-18	UPH9	800–899	$2600	0
Dec-17	402	700–799	$2700	0
Dec-17	212	700–799	$2400	0
Dec-17	503	700–799	$2500	4
Dec-17	528	600–699	$2025	0
Dec-17	LPH4	1200–1399	$4800	0
Nov-17	701	700–799	$2300	2
Nov-17	618	0–499	$1800	0
Nov-17	718	0–499	$1800	4
Oct-17	707	500–599	$2000	4
Oct-17	809	500–599	$1900	2
Sep-17	506	700–799	$2250	21
Aug-17	309	500–599	$1800	5
Aug-17	LPH16	600–699	$2275	15
Aug-17	304	0–499	$1950	1
Aug-17	428	500–599	$1800	2
Aug-17	519	0–499	$1725	0
Jul-17	UPH6	1600–1799	$4950	0
Jul-17	215	600–699	$2050	4
Jul-17	331	700–799	$2400	0
Jul-17	323	700–799	$2600	6
Jul-17	1113	700–799	$2100	6
Jul-17	621	0–499	$1650	6
Jul-17	621	0–499	$1650	6
Jul-17	322	0–499	$1725	0
Jun-17	LPH29	700–799	$2600	0
Jun-17	LPH31	600–699	$1960	2
Jun-17	727	600–699	$2150	8
Jun-17	331	700–799	$2400	0
Jun-17	624	0–499	$,650	1
Jun-17	629	700–799	$2350	0
May-17	317	0–499	$1550	19
May-17	606	700–799	$2100	1
May-17	701	700–799	$2300	10
May-17	301	700–799	$2200	0
May-17	222	0–499	$1650	0
May-17	517	0–499	$1550	0
May-17	418	0–499	$1550	0
May-17	215	600–699	$1895	2
Apr-17	531	700–799	$2400	9
Mar-17	630	700–799	$2400	9
Mar-17	UPH9	800–899	$2550	1
Mar-17	602	700–799	$2450	4
Mar-17	901	700–799	$2100	11
Mar-17	LPH32	1200–1399	$3500	15
Mar-17	514	700–799	$2300	14
Mar-17	601	600–699	$1850	3
Mar-17	713	700–799	$2300	0
Feb-17	708	600–699	$1700	0

Date	Unit	Square Feet	Rent	Days on Market
Feb-17	717	0–499	$1500	0
Feb-17	914	600–699	$2000	11
Feb-17	UPH7	800–899	$2600	13
Feb-17	415	600–699	$1950	0
Feb-17	821	700–799	$2450	0
Feb-17	1015	600–699	$2100	5
Dec-16	910	500–599	$1900	0
Dec-16	1122	700–799	$2500	17
Dec-16	513	700–799	$2300	30
Nov-16	314	700–799	$2300	41
Nov-16	LPH2	1200–1399	$4600	10
Oct-16	UPH7	800–899	$2450	33
Oct-16	411	700–799	$2500	0
Sep-16	519	0–499	$1575	9
Sep-16	1022	700–799	$2500	0
Sep-16	LPH16	600–699	$2150	2
Sep-16	524	0–499	$1550	0
Sep-16	205	0–499	$1700	0
Sep-16	618	0–499	$1550	2
Sep-16	LPH5	1200–1399	$3500	0
Aug-16	1104	800–899	$2600	2
Aug-16	1005	0–499	$1650	2
Jul-16	330	700–799	$2300	9
Jul-16	1102	800–899	$2350	1
Jul-16	LPH7	900–999	$2500	1
Jul-16	621	0–499	$1550	0
Jul-16	LPH1	1200–1399	$4100	40
Jun-16	208	600–699	$1599	9
Jun-16	LPH26	600–699	$2000	0
Jun-16	619	0–499	$1500	5
Jun-16	215	600–699	$1795	30
May-16	222	0–499	$1450	1
May-16	UPH11	800–899	$2450	2
May-16	724	0–499	$1500	3
May-16	UPH8	800–899	$2600	5
May-16	226	500–599	$1750	14
May-16	1011	500–599	$1650	0
May-16	425	1000–1199	$2600	165
Apr-16	617	0–499	$1400	0
Apr-16	819	500–599	$1750	2
Apr-16	1110	500–599	$1650	1
Apr-16	912	500–599	$1650	1
Apr-16	816	600–699	$1850	8
Apr-16	823	500–599	$1700	4
Apr-16	712	700–799	$2050	0
Mar-16	529	700–799	$2300	7
Mar-16	822	700–799	$2100	2
Mar-16	713	700–799	$2250	1
Mar-16	1115	600–699	$1800	2
Mar-16	317	0–499	$1450	0
Feb-16	809	500–599	$1630	1
Feb-16	702	700–799	$2200	0
Feb-16	415	600–699	$1900	3
Jan-16	403	700–799	$2300	48

(Continued)

144 The quantification of risk

Table 6.12 (Continued)

Date	Unit	Square Feet	Rent	Days on Market
Jan-16	516	700–799	$2150	6
Jan-16	721	0–499	$1430	5
Jan-16	629	700–799	$2100	4
Dec-15	616	700–799	$2175	1
Nov-15	912	500–599	$1680	8
Nov-15	1121	700–799	$2175	1
Nov-15	727	600–699	$1950	14
Nov-15	514	700–799	$2200	1
Nov-15	707	500–599	$1700	0
Oct-15	428	500–599	$1625	6
Oct-15	519	0–499	$1450	3
Oct-15	UPH5	1600–1799	$4950	36
Oct-15	531	700–799	$2200	1
Sep-15	708	600–699	$1600	0

Source: condos.ca.

Table 6.13 Expenses at 21 Nelson St.

Monthly maintenance	$100.00
HOA/month	$371.00
Property tax/year	$3050.00
Insurance/year	$600.00
Expense growth	2.7%
CAPEX share of NOI beyond HOA	10%
Sales commission upon sale	5%

Figure 6.12 Canadian dollars per one US dollar

Source: Federal Reserve H10.

Note

1 All figures in the case are quoted in Canadian dollars.

7 When things go wrong

Introduction

In our analysis of property investment thus far, things have looked rather rosy from the perspective of our investor in the 2-tenant industrial building. According to the pro forma we developed in Chapter 5, the investor will earn a 12.26% after-tax return on the expected cash flows on the 2-space industrial building. In Chapter 6, however, we explored variation around this expected outcome and saw that our investor does face meaningful risk beyond what can be reflected in expected cash flows and risk-adjusted discount rates.

In Chapter 7, we explore in more detail circumstances that property investors wish to avoid – bad outcomes where the performance of the property investment is much worse than anticipated. One can imagine many reasons why this may occur. Tenants we have may not renew. Finding new tenants may turn out to be more time consuming and costly in terms of tenant improvements (TIs). Market cap rates may rise. For any of these reasons, our property's return may turn out to be far below what was anticipated in the pro forma. Further, any debt used in the property acquisition will exacerbate the impact of poor property performance on our investor's equity. If we have too much leverage in our investment, we may risk losing the property to our lender.

A closer look at the debt decision

In Chapter 3, we explored how the use of debt influenced the cash flows received by our property investor. At that time, we did not consider how our investor came to decide on the amount of money to borrow. We simply took a lender's willingness to provide credit as an indication that it was reasonable to borrow. In this section, we explore the decision to borrow more closely. The reason is rather straightforward. We are interested in circumstances where the performance of an investment property is significantly worse than what the investor had anticipated. If our property investor did not borrow any money and financed the investment with 100% equity, then poor property performance will matter to nobody but the investor. That is, our investor might lose some money, for instance, if the value of the property declines, but other than that there are really no ramifications of poor performance. The investor continues to own an asset and can work to improve its performance over time. However, an investor that borrows money may not have this luxury. Poor property performance, if sufficiently severe, may lead to our investor not being able to satisfy its debt service commitments to the lender. In such instances, the investor risks losing the property through foreclosure.

146 *When things go wrong*

Therefore, because the use of debt significantly complicates an investor's outcome when a property becomes distressed, it is useful to consider what factors enter into an investor's capital structure decision – that is, how much of a property's purchase price should be borrowed versus supplied by the investor's own equity. In practice, many property investors simply borrow as much as a lender is willing to provide. To the extent that this is true implies that one objective of a property investor seems to be to put as little of one's own equity into the investment as possible. The alternative would simply be to buy a less expensive investment or continue to save until you have the funds to make such an investment.

Investors may also understand the tax deductibility of mortgage interest and use that to justify the use of large amounts of debt when financing investment property. If you've completed the Landlord's Certainty case, you now understand that traditional optimal capital structure theories do not justify the use of debt in (most) real estate investing. This conclusion relates to the fact that personal investment in property is not taxed at a corporate level, but only at the level of the individual investor.

Perhaps most significantly, investors recognize the power of leverage to boost the returns of an investment. In Chapter 4, we calculated that the investor's property delivers a 9.52% return on expected cash flows, but a 16.26% before-tax return to equity once the investor borrows $3,000,000 of the property's $4,600,000 purchase price.

Thus, conserving equity, tax deductibility of mortgage interest, and boosting returns generally motivate an investor's demand for debt. At the same time, lenders appear rather willing to lend against property since real estate generally makes good collateral. As a result, significant debt use is relatively common for property investors. What may be nonetheless underappreciated is that this very same debt use raises the riskiness of the property investment.

Debt use increases risk to property investor

It is useful to illustrate the relationship between debt use and the risk of a property investment with a simple, stylized example.

> Example 7.1: Consider a $1,000,000 property investment over a 1-year horizon. The property has an expected return of 8%, which reflects a 50% chance of the property returning 20% and a 50% chance the property will return -4%. Debt to finance the property is available at 5%. How does debt use impact the expected return and risk of equity investment into the property?

Table 7.1 illustrates the necessary calculations. For example, in the first row of the top panel, we consider the cash flows to the equity investor when the investor does not borrow any of the $1,000,000 purchase price. The property delivers an 8% expected return (perhaps from a 6% cash flow yield and a 2% price appreciation) which flows entirely to the equity investor, who also receives an 8% return. The second row of the top panel illustrates the expected return to the investor when 75% of the purchase price is borrowed at a rate of 5%. The property still expects to deliver $1,080,000 to the investor, who then repays the lender $787,500 = (1 + 5%) × $750,000. The equity investor is then left with $292,500 = $1,080,000 – $787,500, which represents a 17% return on equity.

The bottom 2 panels of Table 7.1 include the actual realizations associated with these expectations. The middle panel demonstrates what happens to equity when the underlying property outperforms expectations. In this scenario, the underlying property delivers a 20% return, which is 12% above its expected value. Equity returns 65%. When the underlying

Table 7.1 Risk and return with leverage

Equity Invested	Money Borrowed	Return on Property	Repayment of Debt	Equity Returned	Return on Equity
		Return of 8%			
$1,000,000	$0	$1,080,000	$0	$1,080,000	8.00%
$250,000	$750,000	$1,080,000	$787,500	$292,500	17.00%
		Return of 20%			
$1,000,000	$0	$1,200,000	$0	$1,200,000	20.00%
$250,000	$750,000	$1,200,000	$787,500	$412,500	65.00%
		Return of −4%			
$1,000,000	$0	$960,000	$0	$960,000	−4.00%
$250,000	$750,000	$960,000	$787,500	$172,500	−31.00%

property does well, investors benefit fantastically from having leveraged their investment. The bottom panel of Table 7.1 considers the opposite outcome. The underlying property delivers a return of -4%, which is 12% below its expected value. In this scenario, the equity investor loses nearly a third of the initial investment!

Thus, debt use makes the equity investment much riskier. In this simple example, we considered deviations of the property's return. of 12% around its expected value of 8%. When borrowing 75% of the purchase price, we see that the return on the equity varies by 48% around its expected value of 17%. If we measure the use of leverage by the ratio of the property value to the equity investment, our investor in this simple example has a leverage ratio of 4 (property value of $1,000,000 and an equity investment of $250,000). This leverage ratio is exactly the extent to which the variability in equity returns varies relative to the unlevered case.

Avoid too much debt

The previous section illustrated in a simple example how the use of leverage increases the risk of the equity investment. The natural corollary to that finding would be for a wise property investor to avoid using too much debt.

How much is too much? This simple question is surprisingly difficult to answer with a great deal of certainty. Let us begin to address this question by appreciating the fact that when developing a pro forma for a potential property investment, more debt will *always* look better than less debt. This is because in our pro forma, it is *always* assumed that the return on the underlying property investment will exceed the explicit interest rate cost of the debt. This mechanically leads to the finding that equity returns will be higher when debt use is higher.

This realization alone suggests that an investor would not want to simply borrow "as much as possible." Rather, an investor should consider the level of risk inherent in the underlying property, and then think carefully about how the use of debt magnifies those risks. Let us return to our example industrial property to examine this process. Recall that in this example, the equity cash flow was expected to be negative in Years 1 and 3 due to vacancy and releasing costs expected at those times. Of course, an investor planning to purchase this building would understand that with debt finance, they will need to be prepared to contribute additional equity into the property in those years above the $4,600,000 that was invested at acquisition. Note that additional equity in Year 3 was only needed because of the existence of a mortgage. The property itself is expected to generate positive (albeit small) cash flow. What if leasing the second space takes additional time, and as a result cash flow in Year 4 is also drastically reduced? Does the investor have sufficient liquidity to contribute even more equity into the project? Or did the investor use the equity it saved by borrowing money to invest in another property, where it cannot be used to finance unexpected liquidity demands of the industrial building?

In another example, suppose that an unforeseen development of competitive properties occurs, and you can only release the first space in the property for $12 instead of the originally planned $18. This change alone reduces the return on the property to under 3%, which is lower than the cost of the debt. If, at the same time, potential property investors view your industrial building to be of greater risk given the new competition, then it may turn out that when you go to sell the property with its lower net operating income (NOI), investors will only value the property at a 10% exit cap rather than the 8% you originally anticipated. This increase in the cap rate results in your after-tax equity returns falling to -9.42%, even though the returns on the property have only fallen to -0.30%.

You may be thinking, perhaps I can simply wait for market conditions to improve. Well, maybe you can't. The mortgage loan you took when you acquired the property had a 5-year maturity. That means that the lender will insist that you repay them the lump sum amount of $2,624,554 at the end of Year 5. Even with the lower NOI and higher exit cap, you would still have enough sales proceeds to repay the lender if you were to sell the property. However, to keep the property you would either need to come up with nearly $3,000,000 yourself or find a new lender to refinance you. In this scenario, however, your property is only worth just over $3,500,000, so it may be difficult for you to be able to borrow nearly $3,000,000 if you were only able to borrow $3,000,000 against the initial purchase price of $4,600,000. Specifically, you would need a lender to provide a loan with over a 74% loan-to-value ratio (LTV) to successfully refinance without contributing any additional equity. Thus, you may be forced to sell when you don't want to (Table 7.2).

So with only a few changes to assumptions (summarized in Table 7.2), this amazing investment proves to be not so amazing after all. And this was the case when the amount of debt used to finance the property was only just above 65% of the initial purchase price. In some instances, and especially in the years preceding the financial crisis of 2008, property investors routinely borrowed 75% or more of the initial purchase price. In some instances, investors borrowed substantially more. In retrospect, it is clear that such high levels of borrowing were not ideal.

Table 7.2 How adversity may force you to sell

Initial Investment Decision	
Initial purchase price	$4,600,000
Initial debt level	$3,000,000
Initial LTV	65.2%
Adverse Outcome	
Assumed rent for Space 1 in Year 3	$240,000
Assumed exit cap rate	10%
Implied Year 5 property value	$3,541,292
Outstanding debt repayment in Year 5	$2,624,554
LTV required to refinance	74.1%

Underwriting standards will work against you

Some investors believe that lenders must know how much would be safe to borrow. That is, if a lender is willing to provide the debt, it must be a good idea. Investors with this belief would simply apply for the largest possible mortgage, and whatever the lender approves would be the choice taken. Of course, this logic fails to appreciate that lenders can make mistakes, too. That is, when property prices are rising, vacancy is falling, and tenants' credit quality appears stellar, banks tend to lend more. What is important for an investor to realize is that underwriting standards change through time, typically following the same economic cycle that influence market prices. That is, when investment properties seem expensive (as indicated by lower than typical cap rates), lender underwriting standards tend to relax as credit generally flows more freely. The converse is also true. At the moment when investors are in the most need for debt finance, its availability is often curtailed.

In Chapter 3, we contemplated the underwriting of an investment property with a value of $2,500,000. We demonstrated how debt service coverage ratios (DSCR), loan-to-value requirements (LTV), and debt yield requirements influence the amount of credit an investor would be able to receive. The first column of Table 7.3 repeats the underwriting choices of Lender A from Chapter 3, only this time let us consider these underwriting criteria to be those found during normal economic times. During such times, one might expect to be able to borrow 62.5% of the value of the property. The second column of Table 7.3 changes the underwriting assumptions consistent with what one may observe during a property boom. During a boom, the lender may look at the tenants and believe that rents will be able to be increased noticeably. Therefore, underwritten NOI increases to $207,500. In addition, the 3 underwriting metrics are set more liberally. It may also be the case that in a property boom, credit spreads on mortgages might narrow so that investors may be able to borrow at a lower interest rate. Further, lenders may allow the loan to amortize more slowly (or not at all in the case of interest-only lending). With these changes to underwriting standards, an investor acquiring the property during a boom may find it relatively easy to borrow 75% of the property's value (or more!).

Table 7.3 Underwriting changes over the business cycle

	Normal	*Boom*	*Downturn*
Property value	$2,500,000.00	$2,500,000.00	$2,500,000.00
Underwritten NOI	$187,500.00	$207,500.00	$167,500.00
Required DSCR	1.6	1.4	1.8
Required LTV	65.00%	75.00%	60.00%
Required debt yield	12.00%	10.50%	13.50%
Interest rate	5.00%	4.50%	5.50%
Amortization period	25	30	25
Mortgage amount	**$1,562,500.00**	**$1,875,000.00**	**$1,240,740.74**
Implied debt service	$109,610.63	$114,004.19	$91,430.80
Implied DSCR	1.71	1.82	1.83
Implied LTV	62.50%	**75.00%**	49.63%
Implied debt yield	**12.00%**	11.07%	**13.50%**
Maximum annual debt service	$117,187.50	$148,214.29	$93,055.56
Maximum monthly debt service	$9765.63	$12,351.19	$7754.63
Maximum loan size from DSCR	$1,670,508.27	$2,437,645.27	$1,262,789.05
Maximum loan size from LTV	$1,625,000.00	$1,875,000.00	$1,500,000.00
Maximum loan size from debt yield	$1,562,500.00	$1,976,190.48	$1,240,740.74

150 When things go wrong

Table 7.4 Underwriting and price changes over the business cycle

	Normal	Boom	Downturn
Property value	$2,500,000.00	$3,000,000.00	$2,000,000.00
Underwritten NOI	$187,500.00	$207,500.00	$167,500.00
Required DSCR	1.6	1.4	1.8
Required LTV	65.00%	75.00%	60.00%
Required debt yield	12.00%	10.50%	13.50%
Interest rate	5.00%	4.50%	5.50%
Amortization period	25	30	25
Mortgage amount	**$1,562,500.00**	**$1,976,190.48**	**$1,200,000.00**
Implied debt service	$109,610.63	$120,156.80	$88,428.60
Implied DSCR	1.71	1.73	1.89
Implied LTV	62.50%	65.87%	**60.00%**
Implied debt yield	**12.00%**	**10.50%**	13.96%
Maximum annual debt service	$117,187.50	$148,214.29	$93,055.56
Maximum monthly debt service	$9765.63	$12,351.19	$7754.63
Maximum loan size from DSCR	$1,670,508.27	$2,437,645.27	$1,262,789.05
Maximum loan size from LTV	$1,625,000.00	$2,250,000.00	$1,200,000.00
Maximum loan size from debt yield	$1,562,500.00	$1,976,190.48	$1,240,740.74

In the third column, I reversed the exercise to simulate borrowing during a downturn. The lender will look unfavorably on the tenants and lower its underwritten NOI. It may adjust its credit metrics to be more conservative while raising its interest rates to address higher perceived risk in the sector. In this scenario, the identical property might only support $1,262,789 in debt, a reduction of over $600,000 relative to the boom for a property valued at only $2,500,000. Thus, changing lending standards can have a dramatic impact on the availability of credit to finance investment property. To some extent, the estimates in Table 7.3 are *understating* the impact that underwriting changes have on the availability of credit to support investment property. This is because the figures in the first three columns are holding the value of the property constant. What if, as may be more realistic, the property's value moves up and down by 20% due to the economic cycle. With this additional assumption, Table 7.4 indicates that the availability of credit may vary by nearly $800,000 over the business cycle for a property that during normal economic conditions would be valued at $2,500,000.

What happens if you default?

The first part of Chapter 7 has tried to convince you that the optimal amount of debt to use on an investment property is less than you probably think and certainly less than what a lender would be willing to provide you in good times. However, it is impossible to foresee every possible bad outcome that your property might face. Further, you may believe that the potential benefits of higher returns outweigh the possible loss of your entire equity investment if things turn out poorly. Despite the careful construction of a pro forma, your property investment may turn out worse than you anticipated, and possibly much worse. Under these conditions, you may not be able to make the monthly payments you promised to your lender or you may not be able to refinance when the loan comes due. What then?

Lender choices

We already know that when a mortgage borrower defaults, the lender has the right to foreclose on the property. **Foreclosure** is the most powerful action a lender can take. By foreclosing, the lender will obtain the title to the property, and at the same time permanently cut off the investor's rights associated with its previous ownership. There are a variety of reasons, however, why a lender may not wish to foreclose. First of all, foreclosures can be both expensive and slow. There are a variety of third-party administrative costs and court costs associated with legal proceedings. Depending on the property location, the proceedings can last anywhere from a month to a year or more. Lenders also worry that during a foreclosure process, the property's value may decline as investors about to lose the property through foreclosure have little incentive to maintain the physical condition of the property or to maintain good relationships with the tenants. If the default happened because property markets were in a downturn, then a lender who forecloses will then end up with a property they likely wish to sell at the worst possible time. This can lead to a firesale discount. Finally, by pursuing a foreclosure, the lender signals to the investor that it is not interested in more amicable solutions. This may tend to increase the likelihood of the investor declaring bankruptcy, which may lengthen the time until a lender can take control of the property.

A foreclosure does not necessarily imply that a lender will recover what it is still owed. In fact, it is generally in situations where the property value has fallen below the outstanding level of debt that a foreclosure happens. Thus, even after taking control and liquidating the property via foreclosure, the lender's obligation may not have been entirely satisfied. If the loan carried recourse, the investor may still owe additional funds to the lender. To pursue its recourse, a lender would then ask a judge for a **deficiency judgment**, which would be the legal right for the lender to pursue the investor for the amount of the outstanding debt that was not recovered through foreclosure.

Note that in a world where foreclosure is immediate and costless, it would always be the best option for the lender. That is because under those conditions, a foreclosure would allow the lender to reclaim the entire value of the collateral property. Thus, the reason lenders consider other options in light of a borrower default is precisely because of the time and expense of foreclosure.

A **deed-in-lieu** of foreclosure is an agreement by which the borrower agrees to deed the property to the lender in exchange for the lender agreeing to cancel all the outstanding debt owed. With a deed-in-lieu, the time and expense of a foreclosure process is avoided by the lender. Also, the borrower may preserve its reputation in the industry by being willing to avoid a foreclosure battle. As a further benefit to a borrower agreeing to a deed-in-lieu, lenders typically forgo the right to obtain a deficiency judgment against the borrower.

Instead of pursuing a foreclosure or deed-in-lieu, a lender may instead attempt to renegotiate the terms of the loan with the borrower in a process known as a **workout**. Since the assumption is that the investor has already defaulted on the existing loan terms, any renegotiation will lead to more borrower-friendly terms. For instance, the lender might agree to lower the interest rate, to forgive (eliminate) some part of the debt, and/or to extend the maturity of the loan. In exchange for offering more borrower-friendly terms, the lender will attempt to extract some concessions from the borrower. These concessions often involve a capital infusion that can offset any deferred maintenance, unpaid property taxes, or expected releasing costs. Note that in pursuing a workout, the lender reserves its rights to ultimately foreclose. Also, it is typical for a workout agreement to contain a lockbox provision (if one doesn't already exist) so that lenders can have more control over the property's cash flow. Workouts might also require the borrower to agree to a deed-in-lieu if there is a re-default on the revised terms of the mortgage.

The lender, however, may choose to forgo a workout and simply accept a **discounted payoff (DPO)**. For instance, suppose the borrower owes its lender $5,000,000, but is unable to repay or refinance. The lender could agree to a DPO by being willing, for example, to accept only $3,000,000 in exchange for a complete satisfaction of loan obligations. Discounted payoffs can typically be negotiated quickly, although lenders must be willing to accept a discount from what they are owed. From the borrower's perspective, a discounted payoff is beneficial because it resolves the mortgage obligation at a lower cost. However, debt reduction is considered taxable income, so borrowers satisfying their mortgage with a DPO will face additional tax liability.

Similar to a DPO is a **note sale**, which means that the lender sells the mortgage note to a different party. From the lender's perspective, a note sale, too, could be quick, yet require a recognition of a sizable loss on the outstanding loan. However, from the property investor's perspective, the note sale is quite different. The borrower would still face the obligation to repay, only this obligation would be to repay a new lender. The note buyer would, of course, have all of the options that were originally available to the original lender.

Note that the different options to resolve a defaulted mortgage loan will likely lead to different recoveries by the lender and will take different amounts of time. Further, the different resolution approaches also entail different risks. Foreclosures lead to a lender bearing all of the inherent risk in being a property owner yet offer the lender the potential to maximize its recovery. On the other extreme, accepting a DPO is likely rather safe, yet necessarily results in the lender accepting a loss.

What influences a lender's choice

When an investor defaults, it is important to understand the incentives of the lender if the borrower desires continued ownership of the property. When a lender considers which option is in its best interest, it will ask itself the following questions.

Why is the loan in default? Of course, the simple answer to this question is that the borrower has not made all of the promised payments on time. When a lender asks this question, however, it is more concerned with whether or not the default was *caused*, in part, by the borrower. Was the default caused by a severe market downtown? Did a large and important tenant unexpectedly declare bankruptcy? Or was the property poorly managed?

How much will each option cost? Is the lender willing to expend the time and cost necessary to pursue a foreclosure? Does the lender have its own expertise to manage a property if it did?

Does the lender have confidence in the borrower? Answering this question goes a long way in helping the lender identify its best strategy. The lender will want to know if the borrower has been forthcoming with all available relevant information regarding the property and its tenants. The lender will only want to renegotiate terms with the borrower if it views the borrower to still be in the best position to manage the property. If the borrower starts filing lawsuits to delay an ultimate foreclosure, it is likely that the lender will not view the borrower as a trustworthy person with whom they would like to negotiate.

When things go wrong: working at workouts

In the Working at Workouts case, you are put in the shoes of Sam Schey, an asset manager at the special servicing firm Drive Property Solutions. Schey's job is to resolve a portfolio of real estate loans in distress, most notably a defaulted mortgage backing a small retail strip mall in suburban Savannah, Georgia. In completing this case, you will be able to:

1 Evaluate and compare the various ways of resolving distressed debt.
2 Recognize and understand how lenders view borrowers in distress.

Working at workouts: commercial real estate debt in distress

Sam Schey, asset manager at Drive Property Solutions, came into his office on Monday, May 10, 2010. He had just returned from a weeklong tour of distressed retail properties in the southeastern United States. Touring commercial properties at various stages of distress was the most fascinating part of Schey's career. His specialty was "special servicing" – the resolution of defaulting commercial real estate loans – a niche industry that had recently become big business in the wake of the severe downturn in commercial real estate.

On his voicemail Schey heard a message from Jonathan Stewart, a lawyer representing Michael Burton, the current owner of Northwinds Community Crossing, one of the distressed properties Schey was overseeing. After deciphering the lawyer-speak, Schey believed that Stewart was offering a financial settlement. With an interest in resolving as many of the distressed properties as possible at the greatest value to Drive and its outside investor partners, Schey anticipated some long days ahead preparing for a protracted negotiation to resolve the problems with Northwinds.

Distressed debt

Traditionally, lenders negotiated bilaterally with a borrower in distress and tried to work out the loan in a way that was mutually beneficial to both borrower and lender. Although the approaches varied from borrower to borrower, workouts typically involved some sort of payment made by the borrower in exchange for some concession on existing loan terms made by the lender. Failing such a workout, lenders had the right to take ownership of the property through the legal process of foreclosure. While within their legal rights, this put bankers in the position of property owners – a less than desirable position in which they had no particular expertise.

Although this approach to managing nonperforming mortgage loans worked well for lenders under normal circumstances, the financial crisis of 2008 and the subsequent dramatic decline in the value of both residential and commercial property led to an unprecedented increase in loan delinquencies. At the extreme, banks burdened with excessive exposure to real estate often failed. As a result, many of the bad real estate loans came to be owned by the Federal Deposit Insurance Company (FDIC), an independent agency of the US government in charge of finding acquirers for the assets and liabilities of failed banks. By May 2010, the FDIC had closed or facilitated the acquisition of 245 banks since the onset of the financial crisis.[1]

As a result, the FDIC had amassed a portfolio of more than $7 billion of largely nonperforming mortgage loans, much of which was secured by commercial real estate properties. The influx of loans was overwhelming and, more often than not, the receivers put in place by the FDIC simply monitored payments while the government agency assembled portfolios of loans to market to bidders for their large note auctions. These auctions were marketed to a dozen or so approved bidders, who would be equity partners in joint-venture relationships with the FDIC for the liquidation of these pools of assets. These investors would assume the risk of nonpayment on the loan in exchange for purchasing the distressed notes at a discount. For example, a bidder might be willing to pay $650,000 for a loan (mortgage note) that carried a face value (promised repayment) of $2,000,000.

Drive Property Solutions

Once investors purchased these distressed notes, their interest was to maximize their financial recovery. Because distressed commercial property was a unique asset, investors often looked to special servicers to try to maximize their recoveries from the nonperforming loans. Drive Property Solutions was a special servicing firm employing 17 people in downtown Chicago. The company's primary function was to resolve distressed debt in the commercial real estate space. Drive's proficiency in resolving nonperforming commercial real estate debt stemmed from its knowledge and expertise in legal remedies (e.g. foreclosure, receiverships), property management should it ultimately own the distressed property, and the working out of distressed loans.

Schey was relatively new at Drive. He had previously worked at a "too big to fail" commercial bank where he monitored regulatory policies and seemed to answer to federal auditors at every turn. Looking for greater career growth, Schey pursued his MBA, focusing on finance and real estate. He would not have predicted his career would turn to restructuring nonperforming debt and foreclosing on delinquent commercial borrowers, but this was the bottom of a deep recession, and Schey felt privileged to have been recruited from a big bank to a thriving new entity such as Drive.

As special servicer, Drive was currently managing a pool of 1000 notes secured by commercial real estate across the country. With the recent boom in distressed commercial properties, however, Drive began investing in the debt as a principal, often in partnerships with private equity firms that supplied most of the capital in exchange for Drive's expertise in dealing with the debt. As such, Drive was one of the few approved bidders for assets auctioned by the FDIC.

With the business of distress booming, Drive recruited Schey as a new asset manager to reconcile distressed debts. Schey was directly responsible for a portfolio of 98 loans that Drive had acquired in the FDIC's most recent auction, involving 64 different borrower relationships. Within a week of joining Drive, Schey had diligently combed through loan documentation, FDIC reports, appraisals, and excerpts of conversations between the FDIC and the borrowers. As he sifted through the documents in each file, he made sure to take careful note of payment history, collateral status, guarantor financials, operating performance, and any other indicative information. Schey then felt ready to focus on his ultimate task – the resolution of these credit relationships with the goal of turning commercial real estate notes into cash.

Property detail

The task at hand was to deal with Northwinds Community Crossing. This property, located at 1701 Northwinds Boulevard in suburban Savannah, Georgia, was a well-maintained 12,209-square-foot retail strip mall situated on 0.97 acres of land. The retail center was configured into 7 tenant spaces, 3 of which had been combined into a single unit.

Table 7.5 Rent roll of Northwinds Community Crossing, May 2010

Suite	Tenant	Sq. Ft.	$/Sq. Ft.	Expiration	Base Rent ($)	Received Rent per Operator ($)
A–C	Mercado Real De La Villa	5160	15.50	9/30/2013	79,980	50,824
D	A&Y Metals	1834	15.50	11/30/2012	28,427	20,725
E	Gerry's Cake Supplies	1849	15.00	11/1/2011	27,735	20,031
F	Tortas Gigantes	1590	16.63	12/31/2012	26,442	18,205
G	Osvelia's Briada	1776	16.00	1/31/2012	28,416	19,236

Table 7.6 Total retail market statistics: Savannah, GA

Period	Existing Inventory # Blds	Existing Inventory Total GLA	Vacancy Total Sq. Ft.	Vacancy Vac (%)	Net Absorption	Quoted Rents ($)
2010 1Q	1,814	21,439,737	1,496,512	7.0	40,645	14.37
2009 4Q	1,812	21,389,393	1,486,813	7.0	47,017	14.91
2009 3Q	1,809	21,334,836	1,479,273	6.9	(55,436)	15.75
2009 2Q	1,808	21,320,016	1,409,017	6.6	(208,782)	16.00
2009 1Q	1,806	21,297,016	1,177,235	5.5	(105,330)	16.54
2008 4Q	1,804	21,279,756	1,054,645	5.0	31,027	16.65
2008 3Q	1,801	21,187,106	993,022	4.7	68,806	16.60
2008 2Q	1,799	21,167,228	1,041,950	4.9	151,553	16.39
2008 1Q	1,798	21,159,858	1,186,133	5.6	61,138	16.32

Source: CoStar.

As he reviewed the building's rent roll, Schey noticed that all tenants had triple net leases, meaning that the tenants paid virtually all property expenses other than property taxes. Somewhat surprisingly, the strip center was 100% occupied despite the severe economic downturn (Table 7.5). Of some concern, however, was the fact that the property's cash flow seemed to be particularly tied to the success of the local community. Schey noticed that the property was located in a predominantly Hispanic market, and that most of the center's tenants catered to that particular market. Due to the collapse in the construction industry as well as stricter immigration laws in Georgia, all of the tenants had suffered a decline in their sales. As a result, Northwinds' current owner, Michael Burton, had allowed his tenants to pay him far less than their contracted rent.

Schey noticed in the file that Northwinds Community Crossing had been appraised in May 2008 for $1,350,000. His friend, Marina McFadden of CB Richard Ellis (CBRE), had recently sent him an opinion of value on the property suggesting that, currently, the property would fetch somewhere in the neighborhood of $805,000 to $906,000, representing sales prices between $66 and $74 per square foot. Schey instinctively knew that CBRE's valuation was only 1 data point. It was based on recent comparable sales in a depressed market and did not reflect the income generation potential of the property. Sam knew that he needed to be confident about the property's worth before beginning his negotiations. This would mean analyzing the data himself, so he began by reviewing the most recent data on the Savannah retail market (Table 7.6). He paused, thinking back to his real estate finance class, where his professor always cautioned, "Skilled financial analysts can make a spreadsheet to justify anything – so think carefully about your assumptions."

When things go wrong

The borrower

The borrower, Burton Properties, LLC, had been formed in early 2003 by Michael Burton for the purpose of operating a convenience store in Atlanta. But after 4 years of running the store and simultaneously building up a reasonable portfolio of similar properties, Burton, a general contractor by training, had changed paths, taking a job at a local high school as the varsity men's basketball coach in his hometown of Savannah.

While he no longer had interest in operating retail outlets, Burton understood the value of real estate holdings as an inflationary hedge for his retirement savings. With the help of his attorney, Jonathan Stewart, Burton sold some of his real estate holdings, and in the spring of 2008 used the proceeds to purchase Northwinds Community Crossing, a property he believed would generate significantly more cash flow and require far less day-to-day management.

The note

In reviewing the loan documentation, Schey realized that the note on Northwinds had followed a circuitous route before landing on his desk. Originally, Burton had arranged financing after contacting his personal banker at Colonial National, describing the property as "a can't miss opportunity." Considering his track record with successful real estate ventures over the previous few years and his longstanding commercial relationship with Colonial, Burton had no trouble securing a commercial mortgage. As Schey looked at the numbers, he was amused by Colonial's loose underwriting (Table 7.7).

Based on the $1,350,000 appraisal, Colonial had been willing to lend $1,250,000 for a whopping 92.6% loan-to-value ratio. With underwriting standards like this, it was no wonder that Colonial had ultimately failed and been taken over by the FDIC in early 2009.

Although the FDIC became his lender, Burton was still obligated to make payments on the note. Before the FDIC's structured note sale, Burton had tried to negotiate lower payments to no avail (Figure 7.1). Drive, in partnership with Spiner Capital, a small private equity shop that had raised money to invest in distressed debt, won the auction that contained the note, effectively paying $465,000 to become the new lender.[2] As was typical of the firm's new partnerships, Drive used its experience as a special servicer and took responsibility for maximizing the recovery on the partnership's investments. Thus, Schey was responsible for its resolution.

Table 7.7 Original underwriting

Original balance:	$1,250,000
Amortization:	20 years
Term:	3 years
Initial interest rate:	7%
Interest type:	After first year, adjusts monthly to WSJ Prime + 2% with a floor of 6% and a ceiling of 9%
Initial payments:	$9691.24/month
Initial annual debt service:	$116,294.88
Maturity date:	May 25, 2011
Default rate:	10% above note rate
Recourse:	Unlimited guarantee of the sole member of the LLC, Michael Burton

In June of 2009, we reached out to the FDIC as receiver for Colonial National Bank (letter attached for reference) to inform them of the then-current position of the property and made a request to approve an interest-only period for 12 months. If the request would have been granted, it would have made it possible to continue timely payments as they were being made at the time. Due to the sale of the asset, we received no response from the FDIC as receiver for Colonial National Bank.

As you will note in the property data provided, all of the tenants are paying a portion of what they are obligated to pay per the lease agreement. Due to the over-abundance of vacant retail space available in the area, a decline in overall sales, and the collapse of the economy, we reluctantly agree to accept a portion of the obligatory monthly rent. If we were not to accept a portion, they would simply vacate and in that case, there would be nothing being collected.

Due to our efforts even in this tough market, we have been able to maintain 100 percent occupancy. The sole reason for this is due to the relationships and trust we have developed with the tenants and the community with our very active, hands-on approach and interaction with each tenant, which requires substantial time and effort.

Figure 7.1 Excerpt of a letter from Burton to Drive Property Solutions, May 31, 2010

Table 7.8 Current status of defaulted note

6 months	$36,082.80
Accrued interest:	$60,138.06
	(default interest)
Target recovery:[1]	$1,250,000.00
Late charges:	$1460.25
Tier:[2]	1
Escrow	$0.00
Total payoff:	$1,300,441.72
Property taxes:	$31,500.44
	(past due for 2010 and 2009; $15,028.69/year)

1 The underlying underwritten value; each note within a structured note sale pool is assigned a value of assumed return (by the equity purchaser) and the aggregate targets are used to determine the bid price to the FDIC.
2 This particular equity purchaser assigned 3 tiers to the assets purchased as thresholds for duration of anticipated hold times (Tier 1 = 1–2 years; Tier 2 = 3–4 years; Tier 3 = 5–7 years).

Schey began his due diligence, determining that Burton had made 18 payments on the original loan, 12 at the original interest rate and 6 at the reduced floor rate of 6%. This had reduced the amount owed to $1,202,760.61. The relationship had since been in default due to nonpayment for 6 months, with the last payment recorded in November 2009. Schey assembled a quick loan summary (Table 7.8) by making calculations based on the loan terms, information from the Spiner/Drive bid at auction, and his contact with the county assessor's office, keeper of the ever-important property tax records.

He noticed that Spiner had targeted a $1,250,000 recovery and expected to resolve this loan within 2 years. Those dollar and timing assumptions nagged at Schey, as he suspected Spiner did not do much formal analysis. Nevertheless, it was a starting point.

Legal options

Schey understood that Drive did not have to negotiate with Burton; the firm could simply take legal action. To be prudent, however, Schey needed to educate himself on the legal environment in Georgia, as laws and practices differed state by state in respect to the 2 most common legal actions taken by lenders following commercial real estate defaults – foreclosure and the appointment of a receiver.

Drive had the legal right to foreclose on the property and take ownership, and Schey understood that in Georgia, this would allow Drive to take ownership in less than 2 months.[3] But he was not sure that Drive would want to do this, as the property was built on a former gas station. The tanks were still in the ground, and while Schey had a Phase II environmental report (Figure 7.2), environmental risks were particularly uncertain. After speaking with Drive's in-house counsel, Patrick Dorn, it was clear that appointment of a receiver was a viable option, executable within a matter of days following filing (Figure 7.3).

In Georgia, however, judges were given the right to appoint the receiver of their choice. Although the lender's recommendation would be taken into consideration, there was no guarantee as to who would be appointed. Through his past experiences, Schey knew that this was a wild card situation. A local judge could enlist a competent property manager to act as receiver who would ultimately work on behalf and in conjunction with Drive. That said, Schey had worked on other cases where judges appointed "friends of the court" to preside over receivership. These individuals, working for the court, could just as easily be incompetent, combative, and ultimately have a negative impact on the property's performance.

It must be noted that prior to conversion to a strip center, the property was a four-pump gas station, demolished in 2006, however, ceasing delivery of gasoline a couple of years prior. A Phase II limited surface investigation was completed by Sierra Piedmont on December 9, 2005, for the subject property. Based upon the findings of their search, Sierra Piedmont did not recommend further assessment at the subject site, stating:

MTBE (0.077 mg/kg) was detected in soil sample B-2 (29–31'). Currently, the State of Georgia does not regulate MTBE. Therefore, this constituent does not need to be reported to Georgia Environmental Protection Division. Since MTBE is soluble and found in gasoline, this constituent typically defines the leading edge of a gasoline plume. However, based on the previous data collected by Sierra Piedmont and the data associated with this report, there is no compelling evidence to suggest that a catastrophic release to soil or groundwater has occurred. Please note that data from both assessments were collected proximal to the existing tank pit. Further assessment will be necessary in the event of tank closure.

The borrower provided the originating bank with a copy of a State of Georgia Notification Data for Underground Storage Tank document dated June 28, 2007. This document indicated that the three underground storage tanks (USTs) on the subject property were protected under the G.U.S.T. Trust Fund. Although this fund has a $10,000 deductible that the owner of the property would be obligated to pay, the G.U.S.T. Trust Fund offers financial protection of up to $1,000,000 due to the former owner (Gas Station 66) making payments to their supplier (Fuel Marketing, Athens, GA), who in turn made Environmental Assurance Fees to the State of Georgia. The storage tanks were installed at the site in February 1991. However, the location has not been receiving gasoline since 2001.

Figure 7.2 Environmental concerns (excerpted from FDIC case filing)

When things go wrong 159

> Receivership is the process of appointment by a court of a receiver to take custody of the property, business, rents, and profits of a party to a lawsuit pending a final decision on disbursement or an agreement that a receiver control the financial receipts of a person who is deeply in debt (insolvent) for the benefit of creditors. Thus, the term "the business is in receivership."
>
> Receivership is an extraordinary remedy, the purpose of which is to preserve property during the time needed to prosecute a lawsuit, if a danger is present that such property will be dissipated or removed from the jurisdiction of the court if a receiver is not appointed. Receivership takes place through a court order and is utilized only in exceptional circumstances and with or without the consent of the owner of the property.
>
> *Georgia Receiverships*
>
> Applicable when the property requires protection, income needs to be collected, and the party in possession is diverting assets or income with the insolvency of the party not being sufficient, alone, as grounds. In this case, due to rental receipts and remittance to the lender below both lease agreements and debt service requirements coupled with the borrower's inability to keep property taxes current, the argument for receivership is clear. It must be noted that the state of Georgia gives the local judge discretion to appoint the receiver he or she sees fit rather than the recommended appointment as sought by petitioning parties.

Figure 7.3 The receivership option

However, without the appointment of a receiver, Schey would never know for sure whether the property's cash flows were as Burton reported. The original note did not contain a lockbox provision, so Burton was collecting the rent personally.

One potential issue with receivership, however, was that if the property was sold through a receivership and the sales proceeds were not enough to cover the principal owed, Drive would lose the right to seek a deficiency judgment against Burton.[4] Given that the loan was issued with recourse,[5] there might be additional value to Drive hidden among Burton's other assets, and Schey did not want to rule out that option either (Table 7.9).[6]

Offer

Looking up from his files, Schey was surprised to see 27 unread e-mails in his mailbox. Deleting the ads for overseas pharmaceuticals and carefully filing the current updates from his fantasy baseball league to read later, Schey spotted a message from Stewart, which had apparently been sent shortly after his voicemail. He read it carefully (Figure 7.4). It came with a number of routine disclosures as attachments – many of which he already had – but one caught Schey's eye immediately: item (iii), the alternative financing commitment letter from First Community Savings Bank (Figure 7.5). He knew he had been hired to bring resolution to these distressed deals and that the obvious way to do this was to have the loan balance paid off in cash. The commitment letter was a boilerplate form, but what Schey noticed immediately was the amount: "the Lesser of $1,000,000 or 75% Loan to Value."

It was clear what was being asked. There was no other direct mention of a discounted payoff, but it seemed clear that Burton was communicating, through his counsel, his intention to request a discount from the principal balance and accrued interest to the maximum $1,000,000 payoff from First CSB. That was some distance from the target given to him by

160 When things go wrong

Table 7.9 Personal financial statement of Michael Burton, May 2010

Assets		Liabilities	
Cash	$3301	Credit Cards	$8409
Marketable securities	$0	Personal lines of credit	$0
Retirement accounts	$295,439	Other	$0
Nonmarketable securities	$0		
Real estate	See below		

Real Estate Holdings of Burton Properties, LLC

Property	Date Acquired	Current Value[a] ($)	Mortgage Balance ($)	Current Monthly Net Operating Income ($)	Current Monthly Debt Service ($)
A	5/14/2003	496,000	525,260	5117	4410
B	8/29/2003	573,000	528,857	6098	4440
C	2/20/2004	1,453,000	1,339,920	13,298	9735
D	4/14/2004	1,089,000	1,364,968	10,003	11,062
E	7/1/2005	161,000	191,041	2224	1787
F	9/2/2005	100,000	87,597	909	745
G	11/27/2006	326,000	322,878	2687	2357
H	4/8/2007	938,000	807,000	7153	5800
I	5/6/2008	162,000	156,000	1419	1500

a Estimated by Michael Burton.

Sam,

My client, Mr. Burton, received your messages. Per your request, attached please find the following documents and information relating to the loan and the property:

(i) Evidence of property insurance;
(ii) Copies of tenant leases;
(iii) Alternative financing commitment letter from First Community Savings Bank;
(iv) History of the property and recent developments in the market;
(v) Statement of assets and liabilities of Burton Properties, LLC.

After you have completed your review of the attached documents and information, please let me know whether there is any additional information that you need and whether you would like to set up a call to discuss our rationale for the loan modification requests set forth in our proposal. We look forward to your response and are hopeful that you will agree that our proposal is a reasonable and practical approach for addressing the issues and concerns regarding this loan and property.

Thank you for your time and consideration,

J. Stewart

Figure 7.4 E-mail from Jonathan Stewart to Sam Schey, May 10, 2010

Drive, but because he had not yet explored what was reasonable, Schey could not dismiss the offer out of hand.

Schey thought, too, about the possibility of a workout. The FDIC had not pursued this course before, but maybe that would be the best way forward. The dimensions of flexibility

Burton Properties, LLC
1175 Magnolia Drive
Norcross, GA 30093

Re: Northwinds Community Crossing

Dear Mr. Burton,

Your application for the first mortgage 03947891Z on the referenced property described further in this letter has been approved by First Community Savings Bank.

Amount: the Lesser of $1,000,000 or 75% Loan to Value
Collateral: Northwinds Community Crossing
Interest Rate: 5.75%
Term: 3 years

Title Insurance: Title Insurance is to be supplied at the borrower's expense insuring the first lien position by a title insurance company acceptable to Lender.

Insurance: The mortgagor will provide all necessary casualty, liability, and flood (if required) insurance in amounts sufficient enough to protect the mortgagee's interest.

Survey: The borrower is to provide a current survey showing all easement of record.

Financial Data: Each year the borrower and guarantor will provide current financial statements by April 15 for the preceding year end.

Prepayment: Loans prepaid at any time prior to the end of the term will incur a penalty equal to 3% (three percent) of the outstanding balance.

Expenses: The borrower is to pay all costs to close the transaction, including the title policy, documentary stamps, intangible taxes, recording fees, lender's counsel, and any other charges necessary to perfect our first lien position.

Other: All loan documents are to be prepared by First CSB's counsel in a form and content acceptable to the lender and will contain all standard loan covenants appropriate to this type of transaction.

No secondary financing will be permitted.

The loan will be due and payable in the event of transfer of ownership through sale of the property or transfer of beneficial ownership in the corporation.

The borrower will provide all necessary representation and warranties as required by First CSB counsel.

Lender will require a Phase 1 environmental inspection confirming that none of the collateral for the loan is in violation of the federal, state, or local environmental laws,

Figure 7.5 Loan commitment letter from First Community Savings Bank

162 *When things go wrong*

> rules, or regulations. In addition, the loan documents will require that the borrower and guarantor will indemnify, defend, and hold harmless the lender of any loss as a result of any past, present, or future use of the property that are in violation of any environmental laws, rules, or regulations and will not permit the property to be used in violation of any such laws, rules, and regulations.
>
> There has been no petition of bankruptcy or reorganization filed by borrower or guarantor.
>
> Lender will hold escrow funds for taxes and insurance.
>
> Value as indicated in this commitment is subject to lender's receipt of a current real estate appraisal by an approved commercial real estate appraiser. The borrower will pay the appraiser directly.
>
> If the above terms and conditions are acceptable please sign and return a copy of this letter no later than May 24, 2010. This loan must close in forty-five days from the date issued or it will become null and void.
>
> Sincerely,
> LENDER
>
> Accepted by:
> Denise Gates
> April 19, 2010

Figure 7.5 (Continued)

were endless, but Schey put together a preliminary proposal as a starting point for any negotiation (Figure 7.6). At the end of the day, perhaps Drive would be better off keeping Burton as the owner, incentivizing him to make payments to Drive in exchange for a lower interest rate, maturity extension, or other relaxations of the existing loan terms. Who knew? Commercial property prices might turn around and Drive could get close to full value for its note in a couple of years.

Of course, a modification that would cause the note to be reclassified as performing would trigger a 35% short-term capital gains tax based on the difference between the modified book value and the $465,000 that Spiner/Drive paid for it at auction. Drive could extend the maturity date up to half the duration of the original maturity without tax implications, but this would only be true if no other terms of the loan were modified. Thus, the avoidance of capital gains tax seemed improbable given that Burton was already failing to meet his current debt service obligations.

Conclusion

As Schey came to the end of his long day, he began typing out an e-mail for a conference call the following week between Drive (represented by Schey and Dorn) and Burton Properties (represented by Burton and Stewart). Schey knew he had a couple of days to fully analyze all of the operating and financial information he had at his disposal. This being his first deal at Drive, he wanted to get the most value he could. Ironically, Northwinds Community Crossing actually showed a lot of promise because it was generating cash flow. The majority of Schey's other loans was collateralized by vacant land or busted development deals.

This Loan Modification Proposal contains an outline of the primary terms of a proposed loan modification. This is not a binding agreement on the part of the Lender, Servicer, Borrower, or Guarantor and remains subject to the negotiation and execution of definitive documentation. This Loan Modification Proposal is for discussion purposes only and sets forth terms that are nonbinding in all respects and may or may not become part of a definitive agreement. This Loan Modification Proposal is not based on an existing agreement between the relevant parties and is not intended to impose any obligation whatsoever on any party. This Loan Modification Proposal is not to be construed as a commitment, offer, agreement-in-principle, or other agreement or understanding of any kind by any party to any term or condition set forth herein or as a waiver by any party of any of its rights or remedies under any agreement, at law, or in equity.

Subject to the foregoing and the other terms and provisions of this Loan Modification Proposal, set forth below for your consideration are the general terms regarding a possible modification of the Burton Properties, LLC ("Borrower") and Michael Burton ("Guarantor") would consider in order to attempt to resolve the outstanding defaults under the loan (the "Loan") currently held by 2009–10 Spiner Capital/Drive Property Solutions/FDIC, LLC ("Lender") and serviced by Drive Property Solutions ("Servicer"):

Past Due Principal and Interest, Interest-Only Period, and Monthly Payment Amount:	Past due principal and interest would be added to the outstanding principal balance of the Loan. The monthly payments would be interest only for a period of six months and thereafter a new monthly payment amount would be set based on such increased principal balance and a twenty-year amortization schedule at an assumed interest rate of 6%. The interest-only period is being requested in order to provide Borrower with a short period to use a portion of the cash flow from the property to help pay a portion of the delinquent real estate taxes. This interest rate is to be fixed for the first year of the agreement.
Delinquent Property Taxes:	Borrower to make monthly payments to the tax commissioner in the amount of $6,000 per month until no tax delinquency exists, provided, however, in the event that Borrower receives any notification of a pending tax sale with respect to the Property, Borrower shall promptly pay all such delinquent property taxes. Borrower shall provide Lender with copies of all tax payments and all tax notices received.
Modification of Maturity Date:	The maturity date of the Loan would be extended to May 25, 2013, in order to provide Borrower with sufficient time to stabilize the property for a refinancing or sale to permit a repayment of the loan without any debt forgiveness.
Discounted Payoff:	Borrower will use good faith efforts to refinance the Loan prior to the maturity date of May 25, 2013, and in the event that Borrower is able to obtain such financing on or before May 25, 2013, then Lender will accept $1,250,000 as payment in full of the Loan.
Modification of Interest Rate:	After the initial year of the agreement, the interest rate would adjust monthly based on the original terms (WSF Prime + 2%), modified to lower the "floor" from 6% to 4%.
Guaranty Release/ Deed-in-Lieu of Foreclosure:	In the event of a default under the loan documents (after giving effect to applicable notice, grace, and/or cure periods) following the execution of the modification agreement, (i) Borrower would either provide Lender with a deed in lieu of foreclosure or permit Lender to pursue an uncontested foreclosure (at Lender's election) and (ii) Borrower and Guarantor would reasonably

Figure 7.6 Loan modification proposal

	cooperate in good faith in connection with the transition of ownership of the Property to Lender and in connection with Lender's ownership and operation, and in exchange Lender shall release Guarantor for any and all liability under the loan documents other than with respect to any hazardous substances, fraud, or misapplication or misappropriation of funds.
Delivery of Financial Statements:	Borrower and Guarantor shall furnish to Lender within thirty (30) days following the end of each calendar year, (i) an annual operating statement and rent roll for the Property, and (ii) a statement of assets and liabilities of the Guarantor.
Cure of Defaults; Notice and Cure Rights:	Upon execution of the modification agreement, (i) the prior defaults regarding past due monthly payment amounts and delinquent property taxes and any other defaults that Lender has actual knowledge of shall be deemed cured and/or waived and (ii) the Borrower shall be entitled to a five-day notice and cure period for monetary defaults (except no notice shall be required with respect to monthly required payments so long as Borrower timely receives a monthly invoice) and a sixty-day notice and cure period with respect to nonmonetary defaults. Any general default regarding a material adverse change in the financial condition of the Borrower, Guarantor, or the Property or regarding Borrower, Guarantor, or the collateral being impaired shall be deleted from the loan documents.
Waiver, Release, and Covenant Not to Sue:	As consideration for Lender agreeing to the above-described terms, the Borrower and Guarantor would agree to waive and release Lender from any and all claims with respect to the Loan and the loan documents through the date of the modification agreement, and would covenant and agree not to sue Lender with respect to any such claims.

Figure 7.6 (Continued)

Notes

1 Federal Deposit Insurance Corporation, "FDIC: Failed Bank List," http://fdic.gov/bank/individual/failed/banklist.html (accessed March 2, 2016).
2 In its winning bid for the portfolio of 98 notes overseen by Schey, Spiner/Drive paid a price approximately equal to 36 cents for each dollar of outstanding loan balance.
3 The foreclosure process in Georgia is completed in 37 days on average, as reported by RealtyTrac, "Foreclosure Laws and Procedures by State," www.realtytrac.com/foreclosure-laws/foreclosure-laws-comparison.asp (accessed September 10, 2012).
4 A deficiency judgment occurs when a court determines that a borrower still owes money to a lender after a foreclosure sale fails to produce sufficient funds to pay off the existing mortgage balance.
5 Recourse means that the borrower is personally liable, above and beyond the value of the real estate, for the failure to pay his or her mortgage debt in full.
6 Schey understood that individual retirement accounts up to $1,000,000 were shielded from creditors if the account owner declared bankruptcy.

Index

1031 exchange 105–110

acquisition fee 88
across-lease risk *see* lease
amortization 50–51
application fee 53
asset management fees 88
assumption 52

balance sheet lender 61–62
beta 27–28
building improvements 23

cancellation option *see* options
capital asset pricing model (CAPM) 27
capital call 85
capital reserve account 24
cap rate 25, 28, 47–48; entry 29; exit 25; going-in 29; going-out 25
commercial mortgage 49–56
concessions 1–2, 4–5, 21
conduit lender 62
credit reduction 21

data table 122–124
debt service 52
debt service coverage ratio (DSCR) 58
debt yield 59
deed-in-lieu 151
defeasance 55–56
deficiency 53; judgment 151
deficiency judgement 151
depreciation 101–102; recapture of 103
depreciation recapture *see* depreciation
direct capitalization method of property valuation 28–30
discounted payoff (DPO) 152

effective net rental income 21–22
effective rent *see* rent
equity accounts 84
exit cap rate *see* cap rate
expansion option *see* options
expense stop 4
extension option *see* options

flat rent *see* rent
foreclosure 49, 53, 145, 151–152

general partners 78
going-out cap rate *see* cap rate
graduated rent *see* rent
gross lease *see* lease
gross potential rental income 20

holding period 19
holding period cash flow 24
hurdle rate 79

indexed rent *see* rent

lease 1–10; gross 3; net 3; present value 7; risk of 8–9; triple net 3
lease present value *see* lease
leasing commissions 24
leasing costs 24
lender yield 56
like-kind exchange *see* 1031 exchange
limited partners 78
loan-to-value ratio (LTV) 58
lockbox 53
lockout provision 54
loss given default (LGD) 57

market rent 20
Monte Carlo analysis 126–133
mortgage 49

net lease *see* lease
net operating income 23
net present value rule 31
note sale 152

opportunity zones 106, 110
options 1, 5; cancellation 5; extension 5; renewal 5
origination fee 53

pari passu 79
percentage rent *see* rent
preferred return 84

Index

prepayment fees 54
price appreciation *see* taxation
probability of default (PD) 57
pro forma 19, 26, 31
promissory note 49
promote 79
property management fee 88

real estate private equity fund 88–90
real estate private equity partnership 78–90
recourse 53
reimbursable expenses 4
renewal option *see* options
rent 1–3; concession 4; effective 7; flat 2; graduated 2; indexed 3; percentage 3
rent concession *see* rent
reversion cash flow 19

scenario analysis 124–126

taxation 100–110; capital gains 103–110; income 100–102; price appreciation 104
tenant improvement 4, 11, 24
total property cash flow 25
triple net lease (NNN) *see* lease

underwriting 49, 56–60

vacancy 2, 8–10; allowance 21–23
vacancy allowance *see* vacancy

waterfall 83–88
within-lease risk *see* lease
workout 151

yield maintenance 54–56